Inventing Transgender Children and Young People

Inventing Transgender Children and Young People

Edited by

Michele Moore and
Heather Brunskell-Evans

Cambridge
Scholars
Publishing

Inventing Transgender Children and Young People

Edited by Michele Moore and Heather Brunskell-Evans

This book first published 2019. The present binding first published 2020.

Cambridge Scholars Publishing

Lady Stephenson Library, Newcastle upon Tyne, NE6 2PA, UK

British Library Cataloguing in Publication Data
A catalogue record for this book is available from the British Library

Copyright © 2020 by Michele Moore, Heather Brunskell-Evans
and contributors

All rights for this book reserved. No part of this book may be reproduced, stored in a retrieval system, or transmitted, in any form or by any means, electronic, mechanical, photocopying, recording or otherwise, without the prior permission of the copyright owner.

ISBN (10): 1-5275-5598-4
ISBN (13): 978-1-5275-5598-3

We dedicate this book to the countless individuals who daren't publicly voice their deep concern about transgendering children, in the hope it will help give them confidence in the future

Table of Contents

Contributors x

Foreword xiv
David Bell

Preface xviii
Gender Critical Dad

Introduction 1
From 'Born in Your Own Body' to 'Invention'
of 'The Transgender Child'
Heather Brunskell-Evans and Michele Moore

Part One: Clinical Perspectives

Chapter One 18
The Tavistock: Inventing 'The Transgender Child'
Heather Brunskell-Evans

Chapter Two 40
Britain's Experiment with Puberty Blockers
Michael Biggs

Chapter Three 56
Transgender Children: The Making of a Modern Hysteria
Lisa Marchiano

Chapter Four 73
Psychiatry and the Ethical Limits of Gender-Affirming Care
Roberto D'Angelo

Chapter Five 93
Gender Development and the Transgendering of Children
Dianna T Kenny

Chapter Six 108
Sex Development: Beyond Binaries, Beyond Spectrums
Nathan Hodson

Chapter Seven 121
Be Careful What You Wish For: Trans-identification and the Evasion
of Psychological Distress
Robert Withers

Part Two: Cultural Perspectives

Chapter Eight 134
'Gender Identity': The Rise of Ideology in the Treatment
and Education of Children and Young People
Stephanie Davies-Arai

Chapter Nine 150
Trans Kids: It's Time to Talk
Stella O'Malley

Chapter Ten 167
Our Voices Our Selves: Amplifying the Voices of Detransitioned Women
Twitter.com/ftmdetransed and twitter.com/radfemjourney

Chapter Eleven 175
Detransition was a Beautiful Process
Patrick

Chapter Twelve 180
Transmission of Transition via YouTube
Elin Lewis

Chapter Thirteen 199
Queering the Curriculum: Creating Gendered Subjectivity
in Resources for Schools
Stephanie Davies-Arai and Susan Matthews

Chapter Fourteen 218
Gender Guides and Workbooks: Understanding the Work
of a New Disciplinary Genre
Susan Matthews

Chapter Fifteen 237
Truths, Narratives and Harms of Rapid Onset Gender Dysphoria
Michele Moore

Contributors

David Bell is a psychoanalyst and Consultant Psychiatrist, Tavistock and Portman NHS Foundation Trust

Michael Biggs is Associate Professor of Sociology and Fellow of St Cross at the University of Oxford. He should be continuing his research on social movements and collective protest but has instead been diverted to defend intellectual freedom and gender nonconformity. He hopes one day to understand how transgenderism so rapidly gained such cultural and institutional power.

Heather Brunskell-Evans is an academic philosopher and social theorist, with a particular interest in the politics of medicine, the sexed body, and the cultural construction of gender. Her current research analyses surrogacy, pornography, prostitution and transgenderism. She is co-founder of the Women's Human Rights Campaign (www.womensdeclaration.com). Heather has published extensively and is a prolific commentator on national and international political agendas driving the rights of women and girls. She hopes that by the time this book is published the scandal of 'transgendering' children will be fully exposed. You can read more of her writing at www.heather-brunskell-evans.co.uk.

Roberto D'Angelo is a psychiatrist and psychoanalyst in a private practice in Sydney, Australia, who works with teens and adults with complex difficulties, including gender dysphoria. He understands all psychological distress to be an emergent property of the person's social and family systems, rather than dysfunction located within the individual. His publications address the limitations of outcome research related to the treatment of gender dysphoria, and the potential that gender transition is not always helpful but can be destructive. He is also a training and supervising analyst at the Institute of Contemporary Psychoanalysis in Los Angeles.

Stephanie Davies-Arai is a communication skills expert, teacher trainer, parent coach and author of the highly regarded book *Communicating with Kids*. She is an experienced speaker on parenting, feminism and 'transgender' children, and has made regular contributions to the House of

Commons. She founded the organisation Transgender Trend in 2015 and has produced comprehensive guidance for supporting gender variant and trans-identified students in schools, for which she was shortlisted for the *John Maddox Prize*, a joint initiative of the charity Sense About Science and the science journal *Nature* in recognition of her work to 'promote sound science and evidence on a matter of public interest, facing difficulty or hostility in doing so'.

Gender Critical Dad is a dad trying to keep his daughter safe. He campaigns and writes a blog at gendercriticaldad.blogspot.co.uk.

Nathan Hodson is a doctor in the UK with experience in surgery and general practice. He holds an Honorary Fellowship in Healthcare Leadership and Management at the University of Leicester. Fertility treatment and reproductive rights are the focus of his research, particularly in relation to the Ottawa Charter's notion that health is a resource for everyday life. He holds qualifications in medical history and medical law and ethics, and has published in leading medical and bioethics journals.

Dianna Kenny is Professor of Psychology at the University of Sydney, Australia. Her cognate disciplines are developmental psychology and developmental psychopathology. She applies developmental neuroscience and psychoanalytic and attachment-informed theorising to issues in infant and child development. Dianna has written extensively on these topics in, for example, *Bringing up baby: The psychoanalytic infant comes of age* (Karnac, 2013) and *Children, sexuality, and child sexual abuse* (Routledge, 2018). For a complete publication listing, go to https://www.researchgate.net/profile/Dianna_Kenny. Dianna is also a practising clinician (see http://www.diannakenny.com.au/).

Elin Lewis (pseudonym) is a Social Studies graduate with over 25 years of experience working within the marketing and digital communications industry. She has specialised in brand messaging, including advising clients on building strong, positive social media relationships. Elin has also worked extensively on marketing and business development projects for the youth and community sector.

Lisa Marchiano LCSW is a licensed clinical social worker and diplomate Jungian analyst in private practice in Philadelphia, USA. Her writings have appeared in *Psychological Perspectives*, *Quillette*, and *Areo*. Lisa has worked with detransitioned young adults as well as with parents of gender dysphoric teens. Lisa's podcast 'This Jungian Life' discusses a range of psychological issues.

Susan Matthews is Honorary Senior Research Fellow in English and Creative Writing at the University of Roehampton. She has published widely on the histories of sex, sexuality and gender. Her research interests include children's literature and narrative theory. Her current research interests include twentieth century narratives of sex change.

Michele Moore is Head of the Centre for Social Justice and Global Responsibility at London South Bank University, Honorary Professor in the School of Health and Social Care at the University of Essex and Editor of the world-leading journal *Disability & Society*. She has worked internationally for more than 30 years, building research-led expertise to support inclusive education and communities. Her work is focussed on consultative participatory and human rights projects across the world to support children, families and those who work with them.

Stella O'Malley is a psychotherapist, writer and public speaker, and is the bestselling author of *Cotton Wool Kids* and *Bully-Proof Kids: Practical tools to help kids grow up confident, resilient and strong*. O'Malley's latest book *Fragile: Why are we feeling more anxious and overwhelmed than ever before?*, released in April 2019, focuses on overcoming stress and anxiety. In 2018, O'Malley was the presenter of the highly-acclaimed Channel 4 (UK) television documentary *Trans Kids: It's Time to Talk*.

Patrick transitioned at the age of 37. After questioning gender-affirming therapy and the transgender ideology, and after experiencing some of the damaging effects of transition, he decided to drop hormone therapy after two and a half years and to share his personal story. He has a YouTube Channel, 'Ein Rückkehrer erzählt', with videos in English and in German.

twitter.com/ftmdetransed grew up as a tomboy and identified as a transman for four years. After an eye-opening meeting with a transman who had transitioned a decade ago and still not found happiness, she felt as if she needed to do more research before going further with her own transition. She soon discovered a major lack of research on the long-term effects of transitioning and also stumbled upon videos by detransitioners on YouTube that completely changed her mind. Ftmdetransed is passionate about amplifying the voices of detransitioners and gives presentations to health care professionals on the topic of detransition.

twitter.com/radfemjourney runs the blog detrans-identified.tumblr.com. Radfemjourney spent years identifying as a transman and as non-binary, and was an activist in the queer community. She started questioning the trans movement after witnessing a known feminist getting no-platformed in 2017 and soon after decided to reidentify as she read more into radical feminism and got to know other desisters. Radfemjourney has since set it as her mission to lift the voices of detransitioned and reidentified girls and women, as she views the representation of this group as a missing puzzle piece in the transgender debate.

Robert Withers is a training analyst with the Society of Analytical Psychology and Senior Lecturer in mind-body medicine formerly at the University of Westminster and currently at the Inter-University College Graz, Austria. In the UK, he is co-founder of the Rock Clinic in Brighton, where he works as a Jungian analyst, psychotherapist and clinical supervisor. He has a long-standing interest in transgender issues and has published several articles on the subject.

Foreword

David Bell

Note: The views expressed here are my own and not those of the Tavistock and Portman NHS Foundation Trust

> 'Every social order creates those character forms which it needs for its own preservation ... The character structure ... is the crystallization of the sociological process of a given epoch' (Wilhelm Reich)

It is great honour to be asked to be asked to contribute the foreword to this book which makes a major contribution to the understanding of the phenomenon of transgenderism and gender dysphoria. The editors have managed to create a work that combines a very high degree of scholarship with a broad reach, including first person moving testimony. It will thus engage a wide readership including academics, clinicians and those parents and families who, like many of us, are struggling to understand this phenomenon.

I have been closely involved in this area for about the last three years, prompted to do so by the sudden exponential growth of children and young people who declare themselves as being in the wrong body, and the pressure for acceptance of this assertion, without sufficient investigation of its basis. Like many, I am acutely aware of the way that proper critical debate has been shut down leaving a near hegemony of a peculiar kind of thinking, or I should say non-thinking, that has come to dominate this discourse. When a movement, can advance through social, political and legal institutions with such a combination of speed and lack of appropriate scrutiny, it is surely right to want to apply the brake in order to create space for critical reflection. It seems to me that over the last year or so we have witnessed the beginning of proper debate, as the media has become a bit more open to a critical perspective, a process in part initiated by the editors' first book on this subject *Transgender Children: Born in Your Own Body* (Brunskell-Evans and Moore, 2018). *Inventing Transgender Children and Young People* builds upon the former work and could not have come at a more important time.

There are multiple routes to Gender Dysphoria which include the presence of various psychological disorders such as depression and autistic spectrum disorder. Then there are children who for multiple and complex reasons live a lonely and isolated life, feeling that they just have no place in the world, and who are psychically lost and homeless. Serious family disturbance is common, often with intergenerational transmission of major trauma such as child abuse in the mother/ maternal line (sometimes a source of the mother's not wanting a girl child). Some families have suffered other major traumas, for example families where the death of a child brings a sibling of opposite sex to seek transition to support an identification with the dead sibling.

A very important causal route, well described in the literature, is related to homosexuality. It is not uncommon for a gay boy, for example, to think that because he is attracted to the same sex he must 'really' be a girl. Some children who show characteristics of being gay/lesbian find this is not tolerated by the family (often very overtly, but equally often in more subtle even unconscious ways); the children internalise this intolerance of their sexual orientation which becomes manifest as hatred of their own sexual bodies. A significant number of these children, if helped in a proper manner would, in all likelihood, end up being gay or lesbian without having undergone transition. This also illustrates the way that gender as a category has come to obscure discussion of sexuality.

We are dealing with a highly complex problem with many causal pathways, and in any case no single causative factor. However, gender services tend towards a damaging simplification. The huge increase in case-loads and long waiting lists lead to pressures to process children using a procedural model rather than one aimed at understanding in any depth the individual case. Of course, alignment with affirmative lobbies (that is lobbies that seek to 'affirm' the wish to change) acts as an ideological support for this simplification.

Many services have championed the use of medical and surgical intervention with nowhere near sufficient attention to the serious, irreversible damage this can cause and with very disturbingly superficial attitudes to the issue of consent in young children. Discussions of the appropriateness of these interventions in children need to be kept entirely distinct from questions of discrimination. The fact that this needs to be stated is, of course, very revealing, as there is an enormous pressure for these two matters to be elided. That is, those who refuse to accept the dominant ideological position and wish to maintain a space for thought and doubt are labelled as 'transphobic', thus serving to silence debate. And this silencing has been remarkably successful, resulting in a simplification of a very complex problem that needs to be understood at *both* individual and socio-cultural levels.

I think it must be made clear that the rapid escalation of referrals, the large increase in natal females seeking to change gender, the sudden appearance of so called 'Rapid Onset Gender Dysphoria', cannot be explained by individual factors alone, nor is it likely to be caused by a large number of individuals feeling free to 'come out' in this new 'liberal' atmosphere. It must be derived from socio-cultural forces which are, as yet, poorly understood, and which need urgent investigation. These might include:

- The penetration of the commodity form into all areas of life so that identity itself comes to manifest features of the commodity. Commodity exchange, because of its extreme rapidity, supports the illusion of instantaneous transformation (I do *not* mean that anyone chooses to change gender without any painful struggle, just that this underlying transformation acts as a tendential force influencing the way we all think).
- I think that in our current conjuncture we are witnessing a growing misogyny. What I have in mind here is this: since the second world war up until the late 70s, strong femininity expressed by the respect for maternal caring, and represented socio-culturally by the welfare state, had a certain degree of social dominance. However, that version of strong caring has been re-presented in its degraded/ perverse form, revealed by such terms as 'nanny state', a contemptuous attack on femininity, reinforced by ideological forms that promote the delusion of the phallic autonomous man, seeking to service only his own needs, enacting a hatred of all forms of dependence. This growing misogyny may be having profound effects on girls which, in conjunction with individual factors, supports the internalisation of this hatred of femininity transformed into a hatred of their female bodies.
- The internet/ social media, a major determining force, occupies a position that is both causal and a vehicle for other causes. Through a kind of viral social contagion, children who feel lost in the world become radicalised on line, join trans groups that provide them at last with an identity, social belonging and an explanation for all their suffering. Further, because of its overwhelming ubiquity and power, it is the medium through which the other factors listed above are transmitted at speed and with no obstruction. This factor is of considerable importance in the very marked increase in the occurrence of so- called 'Rapid Onset Gender Dysphoria', where

onset is sudden sometimes literally from one day to the next. This certainly must underlie localised social contagion, for example in schools.
- Overburdened child mental health services which cannot cope with the combination of increasing demand and radical cutting of resources are stretched to breaking point (Association of Child Psychotherapy, 2018). Faced with children suffering complex serious disorders it is understandable that any mention of gender problems can result in referral to specialist gender services; in the process complex disorders, now filtered through the prism of gender, can be left completely unaddressed. This also leads to a damaging foreclosure of the ordinary turbulence and confusion of adolescence.

It is very regrettable indeed that so many services have sought to treat children *individually*, without any enquiry into this broader determining socio-cultural context. Faced with an individual with a particular disturbance of mind, a psychoanalyst might be asked whether it arises primarily from internal or external factors. This however is the wrong question. All psychological disturbance arises not from inner *or* outer sources but at the point of contact between these two worlds. Gender Dysphoria illustrates perfectly the way a kind of detonating force can erupt where certain inner preoccupations meet explosive outer determinations.

It is an extraordinary achievement of this book that it places itself so firmly on this point of interaction. It will be a key reference work for many years to come

References

Brunskell-Evans, H. and Moore, M. (Eds.) (2018). *Transgender Children and Young People: Born in Your Own Body.* Newcastle upon Tyne: Cambridge Scholars Publishing.

Association of Child Psychotherapists (2018) '*Silent Catastrophe': Responding To The Danger Signs of Children and Young People's Mental Health Services In Trouble.* Retrieved from https://childpsychotherapy.org.uk/sites/default/files/documents/ACP%20SILENT%20CATASTROPHE%20REPORT.pdf

Preface

Gender Critical Dad

I'm 'Bob', more commonly known as a Gender Critical Dad. I wrote a chapter about my daughter deciding she was transgender in the book that came before this one: *Transgender Children and Young People: Born In Your Own Body*. My daughter's school agreed she was trans. Her gender support group agreed. Her friends are so proud to have agreed. Her gay boyfriend agrees. But I don't agree she is really a boy so I wrote about why. *Born in Your Own Body* was a brilliant book, vital for getting people to start questioning the spread of gender ideology in schools and society. I am immensely proud to now be asked to write the Preface for this second book.

When I started to talk about transgender kids in 2016 I felt like a minority of one, then I found others on the internet or in books, reached out on Reddit/GenderCritical and then started my blog gendercriticaldad.blogspot.co.uk. Then, somehow, I was in a room with the other people who were worried: philosophers, other parents, therapists and a wonderful detransitioned woman bursting with life and vitality. I was star-struck. I saw there was a chance to start talking. To be part of a pushback. And now from the first book to a second.

Since *Born in Your Own Body* came out, a new wave of detransitioners and desisters has appeared. With a wisdom and insight incredible in youth, they are unafraid to criticise transgender ideology, narratives and scenes, as they dissect how they were drawn into and manipulated by the trans cult. They are public, determined and getting organised. The detransitioners are finding that coming out as desisting is way harder than coming out as trans. No carnivals, no government- and pharma-funded rah-rah groups, just a long hard look at yourself and negotiating the loss of people you thought were friends.

Kind people ask how my daughter is doing. I can't say too much. It is vital I keep my daughter's identity safe. I have no doubt how activists would hurt her to get at me. So, I have to be vague. I think she's finding her way out of thinking she's trans. I don't ask, because asking would reveal a state of mind, a status to defend. If I close down possibilities for her to keep an open mind, trans lobbyists might win. But these days she wears the binder less, and not at all at home. She had a rant at the poor lad behind the till at the corner shop over the price of tampons and the injustice of the pink tax. All I can really say is that things are calm. She spends term time with a different name and pronouns as far as I know. That's the reality of having her peers, her University, the BBC, the liberal press, all political parties, liberal feminists, and the music and fashion businesses all thinking transgenderism is progressive.

But I think the hold that trans has had is fading. I think change is happening faster in the UK than the US, but the desisters in the US are really coming together and will lead the way for change.

My gut feeling when my daughter first said she was a boy was to keep her away from gender therapists. That was a frightening decision. I know some of her teachers and the parents of her friends thought of me as evil for that. I am glad we did not affirm, glad we did not agree to 'T' and top surgery, that her mum and me did not bottle it. She's safe and well and knows that her parents love her as she is. I protected my daughter from having everyone in her life collaborating in the lie, that she believed, that she was a boy trapped in a girl's body.

A lot has happened. People are waking up to the harms of trans for children and young people, but transgender ideology is still marching into schools disguised as LGBT acceptance. Kids with perfectly normal unease or confusion about bodies and gender are getting dragged onto the transgender conveyer belt of affirmation, blockers, cross hormones, and surgery. LGB kids are passé. Transactivism is a form of bullying: if you're not porn-culture straight, a proper 'boy's boy' or 'girly girl', you must be defective, in need of fixing, by reclassification, drugs or a scalpel. My head is full of questions about why we are allowing this to happen. Is the explosion in the number of transgender-identifying young people a symptom of a society leaving little room for kids to be themselves outside of increasingly restrictive gender roles? How can we keep the conversation going to keep kids safe?

I'm no longer 'the Gender Critical Dad'. I'm just another Gender Critical Dad. There are world-wide networks of parents with transgender-identifying kids supporting each other now. That's a relief. The fight will go on.

Inventing Transgender Children and Young People is a title that promises this book offers some different insights. Its contributors will explain issues, share stories and raise questions that will chillingly illuminate the oppressions, discriminations and hurts spawned from this new invention of 'the transgender child'.

The tide is turning, and I am sure this book will be part of that. If you read it, you will have no excuse for silence.

Introduction

From 'Born in Your Own Body' to 'Invention' of 'The Transgender Child'

Heather Brunskell-Evans and Michele Moore

This book is a sequel to our previous edited collection, *Transgender Children: Born in Your Own Body* (Brunskell-Evans and Moore, 2018a). *Inventing Transgender Children and Young People* extends, develops and strengthens our original aims and intentions, so it is important to reiterate the purpose of the first book.

The aim of *Transgender Children: Born in Your Own Body* was twofold. Firstly, it challenged the concept of a biological basis for transgender identity. The idea that 'transgender identity' is an inherent, biologically-determined phenomenon is not based on well-established, evidence-based principles of medicine, neuroscience, psychology, or psychiatry. From the gender critical perspective of the contributors, we established in that book that the dimorphic sexed body is an empirical fact, whilst gender is the externally imposed set of norms that prescribe and proscribe desirable behaviours for girls and boys. Moreover, the norms of gender are not random, but express and re-enforce gender stereotypes.

Secondly, the book shed light on serious safeguarding issues, both social and medical, which emerge from affirming that children have an inherent 'gender identity'. The book's contributors did not disavow that some children and adolescents experience gender dysphoria and that loving parents will do anything to relieve their children's distress. On the contrary, sociologists, philosophers, psychologists, historians, parents, educators, trans people and de-transitioners collectively acknowledged the current suffering of children and adolescents with regard to gender issues. But the consequences of promoting 'gender identity' as a 'truth' and deploying it in a clinical setting are potentially devastating and life-long. The book called for public debate about the controversial medical practice of treating the healthy bodies of children and adolescents with hormones, one of the effects of which can be sterilisation, on the basis of children's self-affirmed 'transgender identity'.

We began to realise the necessity for a second book to further debate issues previously explored. Three aspects post-publication bore on this decision. Firstly, there was a ferocious attempt to silence the ideas expressed in the book, which included a sustained attack on us editors—on our careers, livelihoods and reputations—the likes of which we had never previously experienced in our long academic careers. Secondly, we became privileged to share conversations and to benefit from the insights of gender clinicians who affirmed the legitimacy of our concerns and privately shared with us the ethical dilemmas they face. Thirdly, many people shared their disquiet about the medical transitioning of children, saying they were afraid to speak out publicly. Gender critical colleagues began to report similar experiences of their ideas being discredited as transphobic. A letter to the Guardian records this collective experience: 'campus protests, calls for dismissal in the press, harassment, foiled plots to bring about dismissal, no-platforming, and attempts to censor academic research and publications' (Guardian, 2018).

Canaries in the mine

Transgender Children: Born in Your Own Body acted as a 'canary in the mine', exposing the fundamentalism that can be attached to the affirmation of 'transgender identity'. In the febrile world of social media, identity politics and knee-jerk emotive judgements, we were prepared for trans activists to attempt to stem the dissemination of the book's ideas. However, the ferocity to shut down the book's oxygen supply was extreme. We enumerate here *some* of the instances of harassment to give a flavour of the powers assumed by trans lobbyists, including academics, in a sustained attempt to discredit us as editors.

Within a few hours of the book appearing on Amazon's website, an organised campaign was underway whereby fake reviews were uploaded that claimed the book to be transphobic and called for it not to be read. After several weeks, Amazon intervened to prevent reviews being written by people who had not purchased the book and removed the fictitious and inflammatory comments. Following the first broadcast discussion of the book by co-editor Dr Heather Brunskell-Evans on the Moral Maze on BBC Radio 4, transgender-identified complainants made an official complaint for her removal from her elected political office in the Women's Equality Party (WEP). The complainants alleged Brunskell-Evans' view that 'gender identity' is socially constructed and not inherent in a child is transphobic, an allegation upheld by the WEP. Days after appearing on the Moral Maze, Brunskell-Evans was no-platformed by medical students belonging to the Reproductive and Sexual Health Society at King's College London where she was Associate Research Fellow. The students claimed that her views on transgenderism 'would violate the student union's 'Safe Space' policy' (Bannerman, 2017).

The *Times Higher Educational Supplement* printed a vexatious book review (Pain, 2018), condemning the book as transphobic. Two rights of reply (Brunskell-Evans and Moore, 2018b; Vigo, 2018) were subsequently published, establishing that the reviewing author was unfamiliar with the contents of the book and had flagrantly misrepresented its aims and purposes. The language of the reviewing author exemplified the current trend in academia to proffer *ad hominem* comments rather than reasoned argumentation whenever contrary views about transgendering children are expressed. Such attacks against academic free speech are contrary to the ordinary reception of critical ideas in academia, where it is normally accepted that disagreement is reasonable and even productive. Some academics now feel free to descend to crude discourse and slurs such as 'TERF' (trans exclusionary radical feminists) to shut down scrutiny of queer perspectives and of the dangers involved in the medicalisation of children and young people. Co-editor Professor Michele Moore withstood a sustained social media campaign comprising false accusations about transphobia and trans-misogyny, calling for her to be removed as Editor in Chief of *Disability & Society* and for a boycott of the journal. Some academics expressly championed Pain's claims that the book was promoting a transphobic discourse and imputed Moore's authority in the world of Disability Studies (Slater and Liddiard, 2018).

How is it that a book advocating a gentle, non-medical approach to safeguarding the bodies and psyches of children could have provoked such an intense attempt to silence and discredit the ideas therein?

Clinicians shine torchlights

During the writing of *Transgender Children: Born in Your Own Body*, we had assumed that clinicians must feel *comfortable* with the practice of transgendering children and must be ethically committed to such work. However, a number of specialist gender clinicians contacted us privately to express their gratitude that the book's gender critical perspective and its rejection of the notion that children could be born 'in the wrong body' opened a much-needed space for a discussion of ideas that were becoming heretical in clinical settings and empowered them to hope for change: 'Thank you so much for all that you have done. Many of us have followed your work closely and have found it inspiring and consoling' (anonymous gender clinician, personal communication, 2018).

We discovered a level of disquiet within gender medicine, including at the heart of the Gender Identity Development Services (GIDS), the main UK specialised gender identity development clinic, based within the Tavistock and Portman Hospital NHS Trust. We were also privy to whistle-blowers from other leading international gender identity services who contacted us to report unrest. It became clear through these exchanges that significant attempts are being made by practitioners, in the UK and other countries, to reach out about their concerns with varying degrees of success and failure. In other words, some clinicians are also functioning as canaries in the mine of transgender ideology and they have deeply informed this book, *Inventing Transgender Children and Young People*.

Behind the public presentation of gender identity development services lies a subterranean stratum of anxiety in some clinicians about the ethics of transitioning children and young people. Clinicians who spoke to us find no evidence for innate 'gender identity', refuting quasi-biological arguments promoted by transgender ideologists such as 'in the first few hours of life or pre-birth, there's a surge of testosterone or something that's made a girl more masculine' (anonymous gender clinician, personal communication, 2018). They expressed concern about being required to work without a cogent evidence-based model for intervention. They described working in an atmosphere in which 'thinking is shut down, questions aren't allowed to be asked and research is never done':

> What I'm saying should be stopped is superficial work, superficial assessment where people are not using their skill set that they've been trained in to think about what is in front of them because they don't have permission to do so. (Anonymous gender clinician, personal communication, 2018)

To summarise, the clinicians who spoke to us identified severe problems within gender medicine which coalesce around the clinical and cultural *invention* of 'the transgender child'. These problems are: an excessively affirmative attitude to the self-identifying transgender child; an inability to stand up to external trans lobby groups; the undermining of a coherent clinical model of child and adolescent development; and serious ethical and safeguarding issues. Clinicians recognise they need to resist the dominant narrative and seemingly intractable 'truth' of 'the transgender child'. They insist it is imperative for groups and organisations inside and outside of gender medicine to challenge the theory and politics of transgendering children, despite the ferocious political backlash it inevitably entails:

> I think that there will have to be a few brave people who put their heads above the parapet. But I think the people who can put their heads above the parapet are those towards the end of their careers who haven't got long to go or much to lose. We need a few of those people to come forward, to say that this is happening and come forward so more and more people will speak out. I think it's already happening. Personally, I think we're seeing the beginnings of a groundswell. (Anonymous clinician, personal communication, 2018)

Clinicians flag up that children and young people are exposed to a range of long-term physical, psychological and social harms because of the inability of gender identity development services to stand up to the pressure from highly politicised campaigners and trans lobby groups who brook no other argument than that the child has been 'born in the wrong body' and demand fast-track, transgender affirmative transition. Further difficulties arise where trans-identified clinicians and others are committed to the values and mission of trans lobby organisations, thus exemplifying that not all clinicians who work within gender identity services share an homogenous model of gender identity and have differing commitments to various outcomes that emerge from treatment.

As this second book was about to go to press, views that had been disclosed to us for over a year were being mirrored in calls to end the 'transgender experiment on children' by five clinicians who had resigned from the GIDS over ethics and safety fears (Bannerman, 2019a).

A year of illumination

The disquiet of clinicians was also being made known within the Tavistock during this period. At the very same time the first book was published, in November 2017, ten whistle-blowing clinicians had contacted David Bell, then staff governor of the Tavistock, about their deep concerns about the practice of transgendering children and young people at the GIDS. By 2018, Bell wrote a report based upon the interviews conducted with the clinicians. He concluded:

> the GIDS service as it now functions [is] not fit for purpose and children's ends are being met in a woeful, inadequate manner and some will live on with the damaging consequences. (Doward, 2019)

Parents of children identifying as transgender, accompanied by co-editor Michele Moore, then of the Patient Safety Academy, Nuffield Department of Surgical Sciences at the University of Oxford, set up a meeting with Paul Jenkins, Chief Executive of the Tavistock, and Dr Sally Hodges, Children, Young Adults and Families Director of the Tavistock. At this meeting the parents handed over a comprehensive research-based portfolio of evidence indicating and substantiating extensive patient safety failings for young people and families in the care of the GIDS. During the same period, articles were beginning to appear in the national press, particularly championed by the *Times*, alerting the public to the potential of a national scandal if the liberal acceptance that medical transitioning is enlightened and progressive cannot be scrutinised.

The Tavistock released a statement in response to media interest about its willingness to engage with whistle-blowers:

> We are disappointed this unsubstantiated report authored by individuals with no expertise in this field made its way to the *Sunday Times* and would urge caution about reproducing its content. It is also important to point out that the report presented hypothetical vignettes rather than actual case studies and does not reflect the practice of the Service. (GIDS, 2019a)

Bell's report was not 'unsubstantiated', since it recorded the evidence of clinicians, nor did it convey the views of 'individuals with no expertise in the field' since all those involved with the report came from the GIDS. The statement that 'the report presented hypothetical vignettes rather than actual case studies' was also untrue since they were based on material provided by clinicians. It is important to note how the Tavistock treats staff who raise concerns, because hostility and misrepresentation will have a chilling effect on the ability of others to speak out.

Following the intensifying of public pressure on the GIDS, the Trust's Medical Director, Dr Dinesh Sinha, was commissioned to produce a GIDS Review Action Plan, which was published in 2019.

It identified some important areas where improvements in the operation of the service could be made but:

> did not identify any immediate issues in relation to patient safety or failings in the overall approach taken by the Service in responding to the needs of the young people and families who access its support. (GIDS, 2019b, p.3)

Sinha's report concluded that 'the Service has sufficient strengths in its area of innovative practice… However, there remains room for improvements' (p.29). Since there are 58 pages of significantly concerning information, it is worrying that Sinha concludes there are 'no immediate issues in relation to patient safety'.

Marcus Evans, a psychoanalyst and one of the governors of the Tavistock, subsequently resigned, accusing its management of having an 'overvalued belief' in the expertise of the GIDS 'which is used to dismiss challenge and examination'. In his email resignation he said:

> In my 40 years of experience in psychiatry, I have learned that dismissing serious concerns about a service or approach is often driven by a defensive wish to prevent painful examination of an 'overvalued system' … I do not believe we understand what is going on in this complex area and the need to adopt an attitude which examines things from different points of view is essential. This is difficult in the current environment as the debate and discussion required is continually being closed down or effectively described as 'transphobic' or in some way prejudicial. (Doward, 2019)

The five previously mentioned clinicians who resigned from the GIDS as a matter of conscience have publicly expressed fury with the GIDS Executive Response (GIDS, 2019b), which stated that it had found no safeguarding concerns. Clinicians say, in regard to the exponential growth in the number of children and young people coming to the service in 2015:

> The whole service should have been halted when the number of 'transgender' cases first exploded. That's the point we should have stopped because we didn't know what we were doing. Are we a service for kids with gender dysphoria, a medical disorder? Or are we a service for 'transgender kids'? (Bannerman, 2019b)

Quite clearly, some clinicians view the GIDS as continuing to be invested in inventing 'the transgender child':

> One clinician said it was understandable if her employer was defensive, saying: 'If they are getting it wrong, you have to ask, are they making kids infertile by mistake? Because if they are to truly acknowledge [our concerns], then they will have to ask themselves, what the f*** have we done to thousands of children? (Bannerman, 2019b)

The invention of the transgender child

Thirty years ago, when gender medicine for children and young people was in its infancy, 'a transgender child' born in the wrong-sexed body would have made no sense to the general public, nor would it have made sense to young people. In the following decades, belief in the existential 'transgender child' has become so universally accepted that it is now counter-intuitive to suggest that 'the transgender child' is an historically invented figure. The mere questioning of whether a boy or a girl can actually be born in the wrong body arouses immense passions in some people, particularly in those who see the practice of transgendering a child as emblematic of a more tolerant, open society. Nevertheless, the contributors to this book demonstrate that 'the transgender child' is not a naturally occurring figure external to current discourses and practices but is brought into being through gender medicine and transactivism. We collectively argue that unquestioning acceptance of 'the transgender child' is unwittingly complicit in the derogation of children's human rights to adult oversight, to bodily integrity, and to have their best interests served.

The book is divided into two parts: clinical and cultural perspectives, as outlined below.

Part One: Clinical Perspectives

In **Chapter One** Heather Brunskell-Evans describes the contribution of the GIDS to the invention of 'the transgender child'. She draws attention to the lack of evidence-based practice, the irreversible harms of hormone therapy, and some clinicians' dissent to practices at the GIDS. She demonstrates that the practice of transgendering children is not progressive and humane but, on the contrary, binds children to traditional gender stereotypes, medically harms them through life-changing irreversible procedures and renders clinicians unable to operate within the medical ethos to which they aspire, namely to 'first do no harm'.

In **Chapter Two** Michael Biggs picks up concerns about hormone therapy to perform a forensic analysis of the GIDS' claim that its service is safe and that there is no evidence of harm. He examines the origins and conduct of a GIDS research project on the administration of puberty blockers and scrutinises the evidence on its outcomes, finding immense flaws in its research design and methodology which have produced misleading conclusions. The research suppresses negative evidence about the deleterious effects of puberty blockers, which calls into question GIDS practices. He concludes there can be no confidence in the GIDS and that the Tavistock Trust has failed not just the scientific community, but more importantly the children in its care.

In **Chapter Three** Lisa Marchiano argues that, in affirming a child as 'transgender', gender therapists induct young people into a feedback loop whereby they classify and interpret their own identity as biologically based. She analyses a current phenomenon whereby medical and mental

health professionals appear to be enthralled by the new condition of 'transgenderism'. A central focus of her chapter is the importance of the unconscious in understanding behaviour. She applies a Jungian analytical model to the unconscious determinants of gender dysphoria, and also to the approach of clinicians who she argues reinforce dysphoria and neglect familial, psychological and social aspects of transgenderism.

Roberto D'Angelo in **Chapter Four** agrees with Marchiano that the biomedical construction of gender dysphoria forecloses the possibility of a 'deeper listening' and leads to a neglect of the multiple situations in which children's gender distress arises out of the complexity of personal and social experience. He explores the ethical limits of transgender-affirming care in psychiatry, suggesting that when children express anxieties about gender they are highlighting, in effect, many problematic social realities, both gender-related and non-gender-related. Perhaps children's current engagement with gender, especially when they perceive it as transgressive or countercultural, reveals a politics about gender for which the affirmation of transgenderism is not appropriate.

In **Chapter Five** Dianna Kenny explores whether gender development in children and adolescents is the product of biological as well as cognitive and social factors. She addresses the complexity of interactions between them and highlights the need for clinicians and early years practitioners working with purported transgender children to have a sensitive and nuanced understanding of developmental stages and processes in order to prevent a precipitous psychological gender transition from which a child may struggle to recover.

In **Chapter Six** Nathan Hodson goes on to examine the politics of how we make meaning of the body. He argues that the proposition that intersex conditions confirm biological sex is socially constructed, as claimed by transgender ideologists, is misconstrued. He explains that intersex is a catch-all term for a range of different conditions, which when examined do not illustrate that we can abandon the material reality of sex, since the division between the sexes remains binary. He concludes that people with atypical sex development should not be exploited for the political purposes of naturalising transgenderism.

Robert Withers draws the specific focus on clinical perspectives on the invention of the transgender child to a close. In **Chapter Seven** he calls upon his own experience as a psychotherapist to argue that affirming gender identity in accordance with the UK Memorandum of Understanding (2017) is not appropriate for helping children and young people resolve gender dysphoria. He warns that complex histories and adolescent confusion over their own possible homosexuality are missed by the memorandum's injunction to avoid 'conversion therapy' by accepting and celebrating every young person's new transgender identity without question. In accordance with other contributors who are therapists, he advocates an approach which does not compel or reject transgender identity but deals with the psychological and emotional factors of gender dysphoria. He says a therapeutic approach is kinder and more facilitative of the personal, physical and social well-being of the child or young person than affirmation.

Part Two: Cultural Perspectives

This section examines the cultural contexts in which the transgender child is invented.

Stephanie Davies-Arai in **Chapter Eight** opens the second part of the book, which turns to cultural perspectives. She describes her work as the Founding Director of the leading organisation Transgender Trend, which pioneers critically concerned oversight about medical transitioning and about the policy and practice of teaching 'gender identity' in schools. She reveals from first-hand evidence the trauma to parents and children caused by the ideology of inherent gender identity promoted in schools and other organisations that supposedly safeguard children, including gender identity services, but which breach children's human rights to bodily integrity.

In **Chapter Nine** Stella O'Malley discusses her work making the television documentary *Trans Kids: It's Time to Talk*. The programme explored her own beliefs as a child that she was a boy, contrasting her own experience with young people with gender dysphoria now. She intended to include contributions from representatives of transgender organisations as well as gender critical viewpoints, but this aspiration quickly saw film production disrupted by arguments and delays as transgender lobbyists sought first to exclude critical commentators, and then to block the completion and broadcast of the programme. Her chapter examines the extraordinary obstacles involved in making a documentary about transgender children and young people. However, she is able to show her project was not entirely one of dismal struggle, since the film provoked a range of extreme reactions that confirm its importance and transformative potential.

In **Chapter Ten** two young women who thought they were transgender and went through with transitioning talk about their experience. Their vulnerability to attack from transgender lobbyists is so real that they cannot write of their experience under their own names. Their chapter tells a story of the difficulties they faced as young women who dared to cross gender boundaries and were then persuaded by transgender activists that transgressions against regressive sex-based stereotypes of femininity proved that they actually were the 'wrong' sex. Their chapter describes a 'psychological rollercoaster' of a journey through transitioning, intoxicating and empowering at first, but which soon descended into doubt, followed by the realisation that they are not transgender, and then having to face a complicated process of detransitioning. They write for the purpose of creating a safer and more inclusive future for women and girls. That they write is, of course, a sign of their power to bring about change to the social heresy of gender normativity.

In **Chapter Eleven** Patrick explores his experience of transitioning, describing how his body and life were subordinated to transgender ideology and the regulation of gender affirming therapy. He describes the difficult years of transitioning, during which his attempts to change his body and identity did not resolve his psychological discontents. Eventually Patrick came to the realisation that he did not need to repress his masculinity in order to be the man that he is. He describes the emergence of a new, more robust rationality about his identity and details the process of detransitioning, which he describes as 'beautiful'. However, in his celebration of detransitioning, Patrick must look over his shoulder and, as with the authors of the previous chapter, cannot write under his full name for fear of reprisals from those wishing to deny that gender is not socially manufactured. Through their courageous accounts, ftmdetransed, radfemjourney and Patrick offer new, powerful representations to make visible the lived experience and consequences of the invention of transgenderism.

In **Chapter Twelve** Elin Lewis looks at how individuals identifying as transgender present themselves as authorities on transition via the popular medium of vlogging, which attracts considerable attention from young audiences across the world. Her analysis shows how unregulated information and advice is given by transgender vloggers through social media in ways which undoubtedly supersede the power of other sources of advice that children and young people may be accessing. The transgender vloggers boast huge numbers of followers which gives them a broad sphere of influence that they use to invent celebrity status images of gender transition, undermining any hesitancies or alternative advice which impressionable viewers may have. She reveals the immense possibilities for harmful influence on young consumers wielded by transgender vloggers. Her chapter highlights that close scrutiny of the influence of transgender vloggers is imperative, given the reach of their messages and the universal lack of control over content.

In **Chapter Thirteen** Stephanie Davies-Arai and Susan Matthews review teaching and learning resources supplied to schools by transgender charitable organisations. Their evaluation pinpoints the ease with which, by supplying free books with stories that will be familiar to young readers and getting children to read them, transgender lobby groups are able to bring transgenderism into a child's world and so begin the process of co-producing the invention of the transgender child. The books bring into being the experience of 'knowing' a child who is 'transgender' and 'being' a child who is 'transgender'. They call for vigilance over the smuggling into schools of curriculum resources which canvass for transgenderism by transgender special interest groups.

In **Chapter Fourteen** Susan Matthews examines learning resources further through her analysis of gender identity workbooks for children and young people which emerged out of queer studies, and which represent transgender identity as innate. These workbooks promise a mythical gender freedom and herald a liberation from gender stereotypes. In contrast, she argues that the workbooks reproduce a litany of gender stereotypes and that, whether addressed to toddlers or adolescents, the goal of the workbooks is to convince the reader of the supreme importance of gender and bolster a new religion of gender identity.

Towards the end of the book the reader will be in no doubt about the invention of 'the transgender child' in the clinical and cultural arenas. In **Chapter Fifteen** Michele Moore addresses the denial of evidence confirming the social construction of transgender children in efforts made to shut down debate about Lisa Littman's evidence of Rapid Onset Gender (Littman, 2018). The integrity of Littman's data destabilises confidence in the clinical diagnosis of transgenderism and exposes deep structural problems for conceptualising any form of actuality of a 'transgender' child. Parents of course, as Littman's data asserted, have been witnessing the need for resistance to transgender ideology for many years, and Moore is able to present new accounts from parents in Scandinavia which open up grave doubts about the invention of the transgender child and the shortcomings of its uncritical intervention bandwagon.

***

We are proud that this collection is made up of chapters from a range of practitioners, academics, cultural influencers and people identifying as transgender and their family representatives. The contributors explore the invention of transgender children and young people through different dimensions of personal and professional experience. This means the book is multifaceted, and the combination of perspectives and internal variability contributes to its strength. The contributions have not been subjected to a co-authored agenda as this would not have enabled justice to be done to the variations in perspective and expertise of the contributors and the different theoretical and experiential tools they each bring to the topic of dismantling the invention of the transgender child.

The chapters address a wide range of contexts in which the transgender child is invented, raising often overlapping concerns which need consideration in a multiplicity of contexts showing *how* the transgender child is constructed by practices in different contexts: from the specialist gender identity development services to resources for schools, gender guides and workbooks; from the world of the YouTube vlogger to the consulting rooms of analysists and psychiatrists; from the pharmaceutical industry to television documentaries; from the developmental models of psychologists to the complexities of intersex medicine; from the voices of clinicians to detransitioners and parents. Far from just documenting the way in which the transgender child is being invented, the authors offer tangible examples of where and how resistance is needed to challenge the infiltration of a dangerous transgender ideology and politics into the minds and bodies of our children and the institutions which are meant to protect them.

By positioning our book between different concerns and across different agendas, we hope to avoid specifically locating our work. We feel the mix of approaches to the topic in the book creates exciting possibilities for new interventional alliances through bringing closer together, across its pages, the experience of those involved in making theory, generating evidence, applying ideas and who are on the receiving end of transgender interventions. Where the non-homogeneity of contributors to the book is completely immaterial, however, is in our shared commitment to raising the critique of contemporary transgender policies and practices. Our refusal to collude in maintaining the claims of transgender lobbyists is unshakeable and we collectively demonstrate this by being part of this writing project.

The book serves three purposes. First, we are offering a resource that can be used both individually and for shared collective development through training by practitioners, clinicians, and others, including ourselves. Second, through the process of producing the book, we are changing ourselves and our activities beyond it; for example, through the various networks, constituencies and affiliations we each have, it is possible to drive change even at the level of conversation. And third, we want to agitate, to put the book out specifically for frictional purposes, to inspire resistance and to create enduring change.

We hope that the book will ultimately deprive transgender lobbyists of the space and opportunity to interject ideology into the schools, communities, clinics, families and other institutions where children must be safeguarded. This time, any 'book burning' campaign will assure us that our critique of transgender-affirmation is powerful. With this second book, we hope there *will* be a backlash, bringing global attention to the public scandal of 'transgendering' children.

References

Bannerman, L. (2017, November 23). Barred academic Heather Brunskell-Evans warns of cowardice over trans issues. *The Times*. Retrieved from https://www.thetimes.co.uk/

—. (2019a, April 8). Calls to end transgender experiment on children. *The Times*. Retrieved from https://www.thetimes.co.uk/

—. (2019b, April 8). It feels like conversion therapy for gay children, say clinicians. *The Times*. Retrieved from https://www.thetimes.co.uk/

Brunskell-Evans, H. and Moore, M. (Eds.) (2018a). *Transgender Children and Young People: Born in Your Own Body.* Newcastle upon Tyne: Cambridge Scholars Publishing.

—. (2018b). Transgenderism, Queer Theory and the Degradation of Academic Debate. *Times Higher Education.* March 22, 2018.

Doward, J. (2019, February 23). Governor of Tavistock Foundation Quits Over Damning Report on Gender Identity Clinic. *The Guardian*. Retrieved from https://www.theguardian.com/

GIDS (2019a). *GIDS review*. Retrieved from http://gids.nhs.uk/news-events/2019-02-17/gids-review

—. (2019b). *GIDS Review Action Plan*. Retrieved from http://tavistockandportman.nhs.uk/about-us/news/stories/gids-action-plan/

Guardian (2018, October 16). Academics Are Being Harassed Over Their Research Into Transgender Issues. *The Guardian.* Retrieved from https://www.theguardian.com/

Littman, L. (2018). Rapid-onset gender dysphoria in adolescents and young adults: A study of parental reports. *PlosOne, 13*(8). doi:10.1371/journal.pone.0202330

Memorandum of Understanding (2017). *Memorandum of Understanding on Conversion Therapy in the UK (Version 2)*. Retrieved from https://www.psychotherapy.org.uk/wp-content/uploads/2017/10/UKCP-Memorandum-of-Understanding-on-Conversion-Therapy-in-the-UK.pdf

Pain, R. (2018, March 15). Transgender Children and Young People: Born In Your Own Body, edited by Heather Brunskell-Evans and Michele Moore. *Times Higher Education.* Retrieved from https://www.timeshighereducation.com

Slater, J. and Liddiard, K. (2018). Why Disability Studies Scholars Must Challenge Transmisogyny and Transphobia. *Canadian Journal of Disability Studies, 7*(2), 83-93.

Vigo, J. (2018, March 22). Body of Evidence. *Times Higher Education.* Retrieved from https://www.timeshighereducation.com/

Part One

Clinical Perspectives

Chapter One

The Tavistock: Inventing 'The Transgender Child'

Heather Brunskell-Evans

In this chapter, I examine the Gender Identity Development Service (GIDS) and its role in the invention of the transgender child. The GIDS is the only specialised National Health Service (NHS) clinic in the UK for children and young people 'who need support for their gender identity' (NHS England, 2015). It is located within the Tavistock and Portman NHS Trust, which remains its institutional home. At the beginning of 2019, the GIDS Multi-disciplinary Team responded to public concern about, amongst other issues, the clinic's administration of puberty blockers and cross-sex hormones to children and young people by explaining the basis on which it administers controversial hormone therapy (see Introduction). Senior members of the GIDS team tell us:

> It is important in the first instance to note that transgender identities have been documented across many different societies and historical time. Nowadays, more and more people are challenging the rigid articulation of sex and gender prescribed by many cultures and voicing an incongruity with their biological sex [...] To account for a strongly felt, unwilled, human capability like gender dysphoria, we probably need multiple-level explanations where the social and the biological intersect. (Butler, Wren and Carmichael, 2019)

To summarise the theoretical principles on which the figure of the transgender child is based, according to the GIDS, these are:

- Transgender identities have existed throughout history
- Transgender identities have been suppressed historically
- It is an example of today's more progressive society that these identities can now be expressed

- 'Transgender identity' is a born property which does not correspond with biological sex

The GIDS team demonstrates a shocking lack of awareness and understanding of the history of ideas that conveys its own affirmative view of transgenderism. The idea that transgenderism is an internal, pre-social phenomenon that has existed throughout history is not an evidenced fact, but a proposition. The GIDS has no credible scientific basis for the theory it applies in a radical and experimental way to children. Whilst it is true that gender expressions are prescribed by history and culture, and that individuals can find stereotyped gender identities painful and restrictive, transgenderism is a newly invented concept. Theories about gender identity were pioneered in the 1960s by sexologists and other academics, and they remain widely contested and poorly understood.

The GIDS offers physically healthy and phenotypically normal children and young people dangerous, off-label drug treatment with life-long deleterious consequences on the basis of a child's subjective feeling, for which there is no scientific test or where clinical diagnosis is based on the child's self-report. In tandem with transgender ideology and political lobby groups the GIDS invents the transgender child, and the impact of this is not only felt by the children and young people who access the clinic's service but by the nation's children from preschool upwards. In the context of the UK legal system which enshrines the myth that transgenderism exists, all children are now being taught that they and their brothers, sisters and friends may have been born in the wrong body (Brunskell-Evans, 2019).

I conclude that, far from safeguarding children confused by gender identity or upholding boys' and girls' human rights to freedom of gender expression and physical and psychological health, the GIDS demonstrates an egregious abrogation of responsibility to protect children and young people from harm.

What is transgenderism?

The GIDS was founded in 1989 by Domenico Di Ceglie and he remained its Director until 2009. At the time of the GIDS' founding in the late twentieth century, the Tavistock had a longstanding commitment 'to psychoanalytic, psychodynamic and systemic theory and practice' which sought 'to understand the unconscious as well as conscious aspects of a person's experience' (Tavistock and Portman NHS Trust, 2019). Its approach to transsexuality (the term still being used in 2002) was that it is a psychological condition for which an individual seeks a physical remedy; individuals believing themselves to be born in the wrong sexed body

> attempt to escape from an intolerable psychological internal conflict into a more comfortable fantasy. Unfortunately, what many patients find is that they are left with a mutilated body, but the internal conflicts remain. (Derman et al, 2002)

In contrast to the therapeutic aim of the Tavistock, which was to help the individual reconcile himself or herself to their biological sex, Di Ceglie innovated an approach which locates the wish to change sex as belonging to 'the area of identity development rather than a specific psychopathological condition' (Di Ceglie, 2018, p.4). His ideas were very much of their time and resonated with concepts formulated in the 1970s and 1980s by transsexuals and trans activists, predominantly in the USA but also in the UK (Brunskell-Evans, 2018). These ideas, formulated before Di Ceglie founded the GIDS, are:

- Gender identity is inherent, an example of human diversity and difference.
- The bifurcation of human beings, like other mammals, into two sexes is not born out by evidence but is a socially constructed imposition assigned to children by society at birth.
- Transgender people are marginalised and oppressed by heteronormativity.
- The suffering of transgender people needs to be alleviated, if they desire it, through medical intervention.

I note there is correspondence between the cultural and political ideas of the end of the twentieth century, Di Ceglie's appropriation of them when founding the GIDS, and the current ideas of senior team leaders described at the beginning of the chapter. We can therefore see that the ideas originated by trans activists were imported into the GIDS by Di Ceglie and are now deeply embedded in the current practices and ethos of the GIDS.

Trouble at the GIDS!

Di Ceglie's ideas were originally disconcerting and rejected by the Tavistock. The main point of contention was that he did not locate gender variance as a psychological condition, as the Tavistock did. He identified beliefs that one has been born in the wrong body 'not as psychiatric conditions but [as] a diverse experience of identity in the gender spectrum' (2018, p.4). He regarded it as more progressive to take an approach where 'altering the perceived gender identity' of children and young people would no longer be 'a therapeutic objective' (2018, p.10). In line with the developing political zeitgeist, Di Ceglie's psychotherapeutic goal was 'to promote well-being and autonomy in the young person ... being sensitive to their needs as opposed to therapy which aims at changing the young person's wishes or beliefs' (2018, p.5). Referencing the Memorandum of Understanding (2017), Di Ceglie says the GIDS objective is 'not prescriptive, addresses the young person's distress and aims at promoting well-being [and] is not part of reparative/conversion therapy and has elements of affirmative therapy' (2018, p.12).

In growing alignment with the ideology and practice of medically transitioning children at the beginning of the twenty-first century, in particular in Holland, Di Ceglie's ideas began to cause consternation at the Tavistock. In 2004, the clinic had already offered puberty blockers to some adolescents around the age of sixteen, however, pressure was mounting from trans affirmative lobby groups to follow the Dutch model of early intervention and allow puberty blockers for pre-pubescent children. Di Ceglie (2018) reminds us there was little empirical evidence of the effects of puberty blockers, and conflicting views of their use pitted (some at) the Tavistock against service users and transgender lobby groups. On the one hand, the Tavistock argued that 'adolescents should only be offered psychological and social support but not the possibility of any physical intervention' (Di Ceglie, 2018, p.8). On the other hand, service users and self-support lobby groups emphasised blockers in early puberty as 'the main way of reducing distress ... and as a life-saving intervention in the form of suicidal and self-harming behaviours' (Di Ceglie, 2018, p.8).

Di Ceglie eventually reconciled these mutually exclusive positions to most stakeholders' satisfaction at that time. As a psychoanalyst, he is bathed in the idea that the psyche speaks in metaphors, and during this moment of tension, he finds a metaphor from Homer's *Odyssey* helpful. Just as Ulysses embarked on an epic journey which involved navigating his ship and sailors between two islands upon each of which sat a monster, one monster threatening to swallow the ship whole and the other to kill the sailors, Di Ceglie, as Captain, navigated the GIDS ship between two opposing groups which, if either had succeeded, threatened the GIDS' very survival. The Tavistock's focus on the psychological aspect of transsexual/transgender identity neglected the material reality of the body;

the trans affirmative lobby groups' focus on the body, including the structure and function of the brain and its influence on gender identity development, neglected the role of the mind. The challenge, as he saw it, was to 'find the middle ground, avoiding the risk of falling foul of either of these two polarities' (Di Ceglie, 2018).

Firstly, Di Ceglie offered the Tavistock the possibility of a research project which would include giving hormone therapy to 'a carefully selected group' of early-pubescent children 'who requested it with the support of their parents' (2018, p.10). The GIDS would then carry out 'a follow-up research project to evaluate the benefits and potential risks' which would give 'accurate, scientific information' (2018, p.10). Di Ceglie tells us 'this project is on-going, and the results are yet to be published' (2018, p.10). Secondly, Di Ceglie decided a model of care needs to take into account the possible biological basis for transgenderism, despite his acknowledgement that neuro-scientific research has consistently attempted but failed to establish a sexed brain (Fine, 2017; Rippon, 2019). He resolved that the GIDS would work in a way which would 'integrate psychological, social and physical aspects in its approach to atypical gender-identity development' (2018, p.10). He placed a core value on *identity* to be at the heart of the GIDS approach to children and young people, shifting the orientation of practice so that the first principle would be for clinicians to affirm the child's self-expressed identity. The balance of power is thus shifted in the therapeutic relationship, away from the clinician's authority and on to the child's beliefs. Di Ceglie says this is important because 'young people are very sensitive and feel intruded upon by anyone attempting to change who they feel they are, in other words *their identity'* (2018, p.10). He assures us that, to this day, the GIDS clinic maintains 'a multi-factorial approach' which has at its core 'an issue of identity' (2018, p.10).

During the same period as the clinic was establishing its specific approach, gender medicine was also being developed outside of the GIDS. The World Professional Association for Transgender Health (WPATH) and the European Professional Association for Transgender Health (EPATH) provide unequivocally trans affirmative clinical guidance and continuing professional development courses for health professionals working with and responsible for the safeguarding of people who identify as trans. WPATH (2019) guidelines on the clinical care of the transgender child and adolescent set out three stages of gender-affirming interventions with progressive levels of irreversibility:

Stage One: Puberty Suppression
Stage Two: Cross-Sex Hormones
Stage Three: Breast, Genital and other Surgeries

A Freedom of Information request revealed that GIDS staff are regularly sent on courses solely provided by these trans affirmative organisations as part of their continuing professional development (FOI, 2018).

Throughout this chapter I explore whether and how this multi-factorial approach that Di Ceglie claims the GIDS operates is realised in practice. I am interested in the following questions:

1. Does the GIDS approach allow for the possibility that a girl, for example, may be mistaken in her subjective, deeply held idea that she is really a boy (and vice versa) and that the most responsible clinical approach to safeguarding her might not be to affirm but help her reconcile with her female body?
2. Does the GIDS approach take into account the influence of social media on children in shaping their self-identity, or take time to therapeutically explore the co-morbidities in the child's family life and personal circumstances?
3. Does the GIDS approach explore evidence-based research to support the view that transgenderism has a biological foundation?

Trans affirmative lobby groups

The Gender Identity Research and Education Society (GIRES) and Mermaids were the major transgender lobby groups to which Di Ceglie was referring in his Odyssey metaphor. It is hard to underestimate the incremental power exerted over the GIDS in the last thirty years by trans affirmative lobby groups. They spearheaded demand for early puberty blockers on the basis that halting puberty would spare children the future trauma of growing into adulthood in 'the wrong body'. The 'wrong body' discourse of GIRES and Mermaids did not arise out of scientific expertise, but rather the founders' personal experiences of parenting their infant sons, the specific meanings they attributed to their sons 'as girls', and their drive to make their own parental meaning-making clinically and popularly accepted as the truth. By 2005, GIRES and Mermaids were so influential that they were invited to a collaborative international symposium of doctors, medical researchers and endocrinologists (Department of Health, 2008). The Department of Health (DoH) discloses that, since this time,

'GIRES and Mermaids remain in close contact with these medical professionals' (Department of Health, 2008, p.3).

The symposium achieved a major success for the transgender lobby in 2009. The Endocrine Society (2009) published new clinical practice guidelines which reduced the age for administering puberty blockers from sixteen to eleven. In 2016, the Government Report on Transgender Equality recommended reducing the amount of time at which cross-sex hormones can be prescribed to below the current age of sixteen (House of Commons Women's Equality Committee, 2016). By 2017, further Endocrine Society guidelines had been issued which, although not substantively different from those of 2009, introduced a conceptual shift in how to think about the gender non-conforming child. The Society now asserts that 'as treatment gains acceptance', the term gender dysphoria should be replaced with the term 'gender incongruence' since the latter term evacuates any imputed pathological dimension to transgender identity but retains the obligation for physical intervention if the child desires it (Endocrine Society, 2017, pp.2-5).

A further outcome of the collaboration between medicine and lobby groups was the creation of DoH guidelines for advising 'gender variant children and young adults and their families' (Department of Health, 2008, p.3). Parents are informed gender dysphoria results from a mismatch between 'the sex assigned at birth' and the child's 'inner sense of knowing they are boys and girls' (Department of Health, 2008, p.4). Despite a lack of neuro-scientific evidence for a sexed-brain, DoH guidelines tell parents the genitalia of a child and the brain can have 'distinctly different characteristics' (Department of Health, 2008, p.4). Parents are advised that acceptance of their child's transgender self-identification is the most appropriate and loving response to what is a 'born' condition. 'Gender identity appears to be indelible from birth' (Department of Health, 2008, p.13). The earlier their child's gender variance is addressed—exemplified by gender stereotypes such as 'exhibiting cross-gender behaviour or dress'—'the more comfortable and happier your child may be as an adult' (Department of Health 2008, p.11).

The guidelines acknowledge that there is a lack of objective criteria for diagnosis, since there is no 'physical test ... for detecting gender variance that may develop into adult dysphoria ... clinicians must rely on the young person's own account' (Department of Health 2008, p.11). They offer numerous website resources for parents and children, including those of GIRES. GIRES designed an e-learning course to be used as a resource for GPs, which the Royal College of General Practitioners (2018) has subsequently withdrawn, the purpose of which was to help GPs respond in their surgeries 'to the needs of adults and young people experiencing gender dysphoria' (GIRES, 2018).

A further contribution to the 'born in the wrong body' discourse arose in 2008 with the development of a third lobby organisation, Gendered Intelligence, which provides workshops for children and adolescents, and provides training courses for teachers and workplaces on transgenderism, inclusion and diversity. The CEO, Jay Stewart, is not a parent, identifies as a transman, and does not have an academic background in psychology but rather in queer theory and the cultural representation of transgender individuals in television (Brunskell-Evans, 2018). Gendered Intelligence insists a medical solution for the young person's gender variance is only one route for those who want the 'inside' to match the 'outside'. It assures adolescents that hormone therapy is not essential, and, in contrast to an approach which focuses on bodies, 'within the trans community we realise it is identity that's more important' (Gendered Intelligence, 2012).

Gendered Intelligence's sexual health guide asserts that retaining one's unaltered sexed body is not an obstacle to transitioning, since what counts is the subjective reality of the young person, their own feeling of being male or female: a male-bodied adolescent who enjoys receiving 'blow-jobs' from a girl is both female and a lesbian if he identifies as such; a female-bodied adolescent who enjoys vaginal penetration is male if she identifies as such (Gendered Intelligence, 2012). The GIDS endorses GI's ideas, advising young people who want to look after their sexual health 'to check out the sexual health guide for trans people from Gendered Intelligence' (GIDS, 2019). A Freedom of Information request revealed that the GIDS 'regularly consults' with GI, as well as with Mermaids and GIRES (FOI, 2018).

The past thirty years have been witness to the invention of two identities for the transgender child: the first is that of the unfortunate victim 'born in the wrong body', i.e. whose gender self-identification requires medical diagnosis and hormone treatment (GIRES, Mermaids); the second is that of the revolutionary adolescent who bravely sensitises the older generation, including trained clinicians, to the subtleties, complexities and politics of gender (Gendered Intelligence). These seemingly contrasting identities are still evolving and taking shape, but are increasingly synthesised into the one figure that we know today, namely 'the transgender child', invested by GIDS with the capacity to consent to hormone therapy and for whom the contradiction by clinicians of their self-identification, for example as truly male in a female body, would be regarded as conversion or reparative therapy.

Myths of GIDS evidence-based practice

Polly Carmichael, the GIDS' second Director between 2009 and the present, has continued with Di Ceglie's original approach. Carmichael was appointed Director in the same year the GIDS was nationally commissioned by NHS England, which set out the specific parameters within which the clinic must operate. An examination of these criteria reveals that gender medicine has developed substantively since 1989 since the GIDS was founded, and now mandates a transgender model of sex and gender identity.

The 2015 criteria for GIDS practice invoke the following foundational principles:

- The GIDS 'will provide specialist assessment, consultation and care, including psychological support and physical treatments, to children and young people to help reduce the distressing feelings of a mismatch between their natal (assigned) sex and their gender identity' (NHS England, 2015).
- The GIDS will 'recognize a wide diversity in sexual and gender identities [and] … support children and young people to understand their gender identity' (NHS England, 2015).
- The GIDS multi-disciplinary team will include 'paediatric and adolescent endocrinologists', thus validating the hormone therapy that was initially so problematic (NHS England, 2015).

The GIDS advises children that the criteria for their eligibility to consent for hormone treatment includes: 'you have a strong feeling that you are in the wrong body or that your gender and body do not match'. The possible benefits of starting on hormone blockers are the following:

- They may improve the development of your body in your desired gender
- If you decide to stop hormone blockers early your physical development will return as usual in *biological gender* [sic, my italics]
- The hormone blockers will not harm your physical or psychological development
- Hormone blockers will make you feel less worried about growing up in the wrong body and will give you more time and space to think about your gender identity

- Hormone blockers may reduce the amount of operations that you may need as an adult (after the age of 18) should you wish to have operations to change your body (GIDS, 2019)

Carmichael asserts this model of care is based on 'a practical, evidence-based approach' (Butler et al, 2018, p.631). This claim is both misleading and erroneous. Firstly, the GIDS team acknowledges 'much is still to be ascertained about the impact of medical, intervention' and that 'the long-term physical effects of chemically delaying puberty are virtually unknown' (Butler et al, 2018, p.636). Information given to children when ascertaining competence for consent says:

> We do not know how hormone blockers will affect bone strength, the development of your sexual organs, shape or your final adult height [...] and other long-term effects of hormone treatment in early puberty that we don't know about [...]. (FOI, 2019)

Secondly, the GIDS withholds from children evidence that hormone blockers do not necessarily provide psychological relief for gender dysphoria. Although Carmichael assures children they will experience relief from anxiety, she elsewhere acknowledges research evidence that, after one year, young people report an *increase in internalising problems and body dissatisfaction* (Carmichael et al, 2016). The GIDS team admits that puberty blockers will almost inevitably lead to cross-sex hormones which will invariably lead to sterility (Butler et al, 2018).

Thirdly, it is difficult to produce evidence-based research because gender identity is a subjective, not objective, truth. The GIDS team agrees it has no understanding of the aetiology of transgenderism or how to understand its exponential growth (Butler et al, 2018), but nonetheless holds steadfast to the proposition that 'evidence from twin studies and brain differences, although tentative, suggests at least in part a biological component to gender diversity (Butler, Wren, and Carmichael, 2019). No solid scientific evidence is ever produced by the GIDS which could substantiate this claim; it is a proposition whose veracity is to be taken on faith, something one day we *may* discover as fact. However, what we do know is that inherent, pre-social gender identity is not a scientifically established fact but a disputed concept with a political history, and that the GIDS practice is far from evidence-based.

The endocrinologist Michael Laidlaw, critical of hormone therapy, points out that gender identity, let alone transgender identity, is not a physical phenomenon like sex that can be located in the body. He says: 'there are no laboratory, imaging or other objective tests to diagnose a "true transgender" child' (Laidlaw et al, 2018). Laidlaw points out that gender dysphoria is 'not an endocrine condition but becomes one through … puberty blockers and high dose cross-sex hormones' (Laidlaw et al, 2018). He reminds us that the practice of affirming 'gender identity' through hormone therapies is extremely serious, including: 'potential sterility, sexual dysfunction, thromboembolic and cardiovascular disease and malignancy' (Laidlaw et al, 2018). Hormone blocking means the gonads never mature, and immature gonads with immature gametes are never fertile. The drugs which block hormones (GnRHa) have not been certified as safe by drug manufacturers, and the physical and medical impact must be measured against the fact that children with gender dysphoria will outgrow this condition in 61-98% of cases (Laidlaw et al, 2018). Laidlaw concludes that in effect what adolescents have 'bought' themselves is not time but 'simply, lower bone density and the need for lifelong medical therapy' (Laidlaw et al, 2018). The health consequences of gender affirming therapy through puberty blockers and cross-sex hormones 'are highly detrimental, the stated quality of evidence in the guidelines is low, and diagnostic certainty is poor' (Laidlaw et al, 2018).

The Endocrine Society's 2017 guidelines confirm that 'most children who have gender dysphoria actually lose it … There may be only 10% to 15% whose dysphoria continues throughout childhood and into puberty' (Endocrine Society, 2017). Researchers carried out a study of puberty blockers in adolescents between the ages of 11-17 which provides evidence that hormone blockers promote a continued desire to identify with non-birth sex—over 90% of young people attending endocrinology clinics for puberty-blocking intervention proceed to cross-sex hormone therapy (De Vries et al, 2018).

Christopher Richards, a consultant paediatrician, draws attention to much empirical evidence that the use of puberty blockers leaves a young person 'in developmental limbo without the benefit pubertal hormones or secondary sexual characteristics' (Richards et al, 2018). He points out that hormone therapies are being used 'in the context of profound scientific ignorance' (Richards et al, 2018). He argues that 'to halt the natural process of puberty is an intervention of momentous proportions with life-long medical, psychological and emotional implications […] this De Vries practice should be curtailed until we are able to apply the same scientific rigour that is demanded of other medical interventions' (Richards et al, 2018).

Carl Heneghan, Professor of Evidence-Based Medicine at the University of Oxford, carried out an analysis of the research studies about hormone therapy and concludes that there are 'significant problems with how the evidence for puberty blockers and gender-affirming cross-sex hormone has been collected and analysed that prevents definitive conclusions to be drawn' (Heneghan and Jefferson, 2019):

> The evidence is limited by small sample sizes; retrospective methods, and loss of considerable numbers of patients in the follow-up period. The majority of studies also lack a control group (only two studies used controls). Interventions have heterogeneous treatment regimens complicating comparisons between studies. Also, adherence to the interventions is either not reported or inconsistent. Subjective outcomes, which are highly prevalent in the studies, are also prone to bias due to lack of blinding.

Heneghan reminds us that treatments for under 18 gender dysphoric children and adolescents remain 'largely experimental' (Heneghan and Jefferson, 2019). There is ignorance of the long-term safety profiles of the different hormone regimes. The large number of unanswered questions include: the age at the start, reversibility, adverse long-term effects on mental health, quality of life, bone mineral density, osteoporosis in later life and cognition. He concludes: 'The current evidence base does not support informed decision making and safe practice' (Heneghan and Jefferson, 2019).

Clinicians 'working at the edge'

Di Ceglie acknowledges that GIDS clinicians often experience a particular kind of pressure and anxiety: they may learn through personal discomfort where they are positioned, at either 'the edge' or on 'the mainland' of the GIDS' ethos (Di Ceglie, 2018, p.6). Here are some of the views, both spoken and written, recorded by myself and my co-editor, Michele Moore, in conversation with a number of GIDS clinicians who feel they are 'at the edge'.

An anonymous clinician says:

> I can imagine that it felt important historically to people who were there at the beginning [of the GIDS], early on when this was a very rare and unusual thing, and people were … finally getting the service up and running through a very circuitous route … perhaps there was a need to eschew some of those [psychological] models [of child development] because in those days affirming … you know affirming would have meant listening and hearing and understanding, it wouldn't have meant this over-determined word … perhaps it really needed to be a maverick service.

Another anonymous gender clinician (personal communication, 2018) posits that many clinicians are now subject to 'the trans zeitgeist and internal advice rather than on our own knowledge of child development, mental health or any kind of evidence base':

> The invitation is to forget all we know if that makes sense ... so I guess what I think should be stopped is this practice where you don't use your own training, your own models of making sense of child development ... it feels like an injustice, and it feels like a responsibility because there's so few people witnessing this. All the undergraduate stuff about ability to think abstractly, about the adolescent brain, you know the risk-taking, all of that stuff we use in all our work, everywhere we're working with kids, that 'truth', we're not using in this service.

One anonymous clinician says that adolescents' uncomfortable feelings about their sexually maturing body are normal, but these are being classified as the young person having been 'born in the wrong body'. It is clear that, for children identifying as transgender, there are co-occurring social and psychological difficulties, and without a cogent model for why a child is presenting with gender dysphoria other than gender identity affirmation, clinicians struggle to formulate a diagnosis and intervention plans which take into account variables such as trauma, difficulty, distress, and psychopathology. The trans affirmative model 'precludes formulation in the classical sense'. According to an anonymous gender clinician:

> When the kids have been the victims of sex crimes, misogyny, rape, sexual abuse and/or are on the autistic spectrum, clinicians are either not acknowledging the possibility of these factors being related to gender dysphoria and/or are simply not seeking out any more knowledge that might tell them so. This obviously sets up a confirmation bias where children are only able to be been seen through the 'T' lens.

On being questioned about why trained, qualified clinicians capitulate to affirmation, another anonymous clinician says they have 'witnessed something of a process occurring in many staff which goes some way to explain this': staff who arrive from other services are struck by a sense of entering into a completely different culture with different approaches to other psychological services within the NHS. There are a few stages that clinicians go through after they enter:

1) Naivety—where you do as you are told, essentially, as you feel that you are totally deskilled and simply do not yet have the secret knowledge that must exist;

2) A bit of an awakening—where you start to question things and become a little distressed;
3) A time when you speak out internally—this helps people to feel like they are not just capitulating but are challenging things. This stage is seductive and often allows people to continue doing the work for some time under the misguided (I believe) belief that they are actually making the situation better;
4) A more depressive/despairing stage where people realise that change has not been possible and that they are part of the system that is continuing to take risks with kids. Some people tend to bounce between stage 3 and 4 for a bit until they leave. The people who stay are either in stage 1, 2 or 3, I think—they have perhaps not allowed themselves to reach stage 4.

Another group simply have an entirely different belief system which is guiding their work, so they do not go through any of this because they are happy with self-affirmation.

Clinicians expressed anxiety to us about the power wielded by some parents and the background support of those parents by trans-affirmative clinicians:

> Some families have a lot of power, there's a sense they're going to fight to get to the top of the service in order for you not to be able to get in their way. The way that power operates is that there are very strong powerful voices of clinicians prevailing *within* the service who have a very particular way they believe we should be moving kids through ... A lot of the good work that does not go along with this [is] quite subversive or underground, with the support of colleagues who have shared views, but where it does feel like you're not being backed by the service – which is terrifying. (Anonymous gender clinician, personal communication, 2018)

Another anonymous clinician (2018) says: 'I think for clinicians it's that toxic combination where it's not only in this context, but the wider society too, that even going down this route of thinking critically is seen as being completely unacceptable, even to be thinking about it, questioning it'.

The clinicians who resist diagnosing the gender dysphoric child as having an innate transgender identity also 'risk being accused of transphobia and practising conversion therapy' (anonymous gender clinician, personal communication, 2018). For many GIDS clinicians:

> There are elephants everywhere [at the GIDS] and what is so shocking is that they are not acknowledged, even by seemingly critical people who in other forums would be critical and individual in their thinking.

> Affirmation of gender identity is really powerful ... it's not just in the service, it's happening all over, I mean this cultural phenomenon which is happening in the west, isn't it, we just see it undiluted.

> The total lack of evidence-base for anything we are doing, the lack of permission to use any kind of explanatory model for what you see, so a formulation, or a hypothesis, we have no permission to make one.

> In our service, by not acting on our thoughts and not going the next step in terms of questions, the outcome can be a child who is not able to have children—which is so profound. It's such a profound responsibility.

Sex and 'gender identity': constructions and contingencies

In this chapter I have demonstrated that the GIDS has committed itself to only one model of the distinction between sex and gender, namely the transgender affirmative model such that transgender ideas of sex and gender have become central to medical practice at the GIDS. However, it is the GIDS' commitment to the transgender model of sex and gender which is bringing the Service into ethical disrepute.

There are alternative models that support the conceptual distinction between sex and gender, most notably a gender critical model, which the GIDS does not explore. The gender critical model refutes the idea that gender is inherent, and demonstrates that unlike biological sex, which is a material reality and not socially constructed, gender is socially produced. I acknowledge that all models of the relationship between sex and gender are constructs, and a gender critical model, like the gender affirmative model, is an interpretation. However, the gender critical model has many ethical and epistemic advantages. Firstly, it is not based on tentative, unverifiable claims about the relationship between the brain and inherent gender, but on the unassailable evidence that gender is specific to our own time and culture, and that 'femininity' and 'masculinity' realise themselves in different social contexts and historical periods—an approach which it may be both difficult but also valuable for a young person to understand. Secondly, it supports young people to be comfortable in their own bodies: a gender critical model does not confine boys and girls to gender stereotypes of 'masculinity' and 'femininity' such that, for example, a boy attracted to stereotypically 'feminine' activities or clothes may 'truly' be a girl trapped in 'the wrong body'. A gender critical model helps avoid a lifetime of medical intervention with life-long deleterious consequences which will not, despite the young person's possible fantasy, ultimately transform the body to the opposite sex.

In contrast to the trans claim that transgender children have always existed, Susan Matthews argues the figure of the transgender child is a product of 'our moment in history, which invents a new set of beliefs that are without historical precedent' (Matthews, 2018, p.135). Transgender doctrine 'relies centrally on an appeal to personal belief coupled with an emotional appeal to protect the oppressed … it is the product of a post-truth world' (Matthews, 2018, p.135). The transgender child is not a pre-discursive figure awaiting discovery, rather it is a recent invention: '"the authentic self", realized through "gender affirmation" is as historically new as the technologies that make it possible' (Matthews, 2018, p.124).

Michele Moore (2018) demonstrates that the transgender child is iatrogenic, occurring not only against a backdrop of social media grooming and the increasing role of trans affirmative ideas in educational provisions for children and their teachers and parents, but also the trans affirmative approach in gender medicine itself which assumes, though has not demonstrated, a physical foundation to transgenderism. She says:

> The idea that the cause of 'misaligned' gender lies in a child's body is central to transgenderism, so that medical intervention can be countenanced even though the inevitable result of such interventions will be a life-long pursuit of difficult and painful physical and psychological transitioning that will uphold and deepen socially constructed gender-based oppression and never actually change a person's biological sex. (Moore, 2018, p.220)

Moore points out that there is a circularity to the logic of transgender doctrine, which makes inescapable the view that transgenderism is 'a self-fulfilling prophecy':

> Children self-identify as transgender, social and medical intervention takes place thus confirming and intensifying the child's self-diagnosis; parents, teachers, other professionals and care-givers are induced to understand gender non-conformity as evidence of transgenderism, to accept transgender doctrine as 'truth' and to collude with social and medical intervention. (Moore, 2018, p.220)

Bernadette Wren, Consultant Clinical Psychologist at the GIDS and the Tavistock Head of Psychology, straddles two uncomfortable, incompatible positions with regard to transgenderism. On the one hand, as said at the start of the chapter, she asserts that 'transgender identities have been documented across many different societies and historical time' (Butler, Wren and Carmichael, 2019). On the other hand, she acknowledges the value of the gender critical perspective to the extent she agrees that 'the meaning of trans rests on no demonstrable foundational truths but is constantly being shaped and re-shaped in our social world' (Wren, 2014, p.271). She accepts the GIDS is 'in the business of helping actively construct the idea and the understanding of transgender' (Wren, 2014, p.284). It is not just transgender identity which is socially constructed, but to a certain extent all identities. She recognises 'the constructed, contingent, unstable and heterogeneous character of subjectivity, of social relations, of gender and of knowledge' (Wren, 2014: 285).

Wren acknowledges that different clinicians as well as non-clinical groups within society occupy different discursive spaces about the wisdom of transgendering children, differences that 'militate so powerfully against developing shared understandings of what is ethical and compassionate practice, what 'good' care looks like and who should have a definitive say in treatment decisions' (Wren, 2018). She tells us that 'conflicts in the field of gender identity and the ethical care of children and young people relate as much to the very possibility of making meaning, the right to make meaning and the appropriate focus for meaning making' (Wren, 2018). She states that the work of the GIDS sits 'at the intersection of a range of deeply contested and radically unresolved (possibly unresolvable) social and cultural issues of our time'. She asks: 'How can we go on with no foundational beliefs in which to take refuge? She answers this by saying that the clinician's work is 'to take a working stance ... when young trans clients make demands for physical intervention' (Wren, 2014, p.284). She asks a question of the GIDS which I too would like to pose: 'How do we justify supporting trans youngsters to move towards treatment involving irreversible physical change, while ascribing to highly tentative and provisional accounts of how we come to identify and live as gendered?' (Wren, 2014, p.271).

Wren concludes that the GIDS' decision to recommend hormone treatment for any individual young person is 'a genuinely shared but imperfect decision, involving the client, family, other professionals in the context of a wider cultural world, in which the meaning of trans is constantly shaped and re-shaped, but which rests on no foundation of truth' (Wren, 2014, p.287). She invokes uncertainty about gender and knowledge to salve the clinician's anxiety, and, to an extent, to let herself off the hook: responsibility for constructing and offering hormone therapy is 'shared'. Meanwhile sexed bodies hardly exist. It is the mutable meaning of transgenderism that is Wren's focus. The GIDS' decision-making is 'imperfect'—this is just how things are in the current so context. Wren could equally have invoked a completely stance for the GIDS therapist, whilst still retaining poststructuralist theoretical model. She could have argu... modern individual is thoroughly invested with cultural discourses about sex and gender such that they experience their identity as pre-social, then the ethical stance for the GIDS should be a *refusal* of the affirmative model of 'the transgender child'.

The ethical framework in which the clinic operates in order to ascertain whether a child has the competence to consent is a model for correct practice and procedure, not an overall consideration of the ethics of hormone therapy. A child cannot truly give informed consent to hormone therapy for the following reasons:

- The medical consequences are extremely complex, and a child will have little or no cognisance of a future in which he or she may come to regret lost fertility or the lack of organs for sexual pleasure.
- In contrast to the staggeringly naive GIDS proposition that the child can give consent if he or she has been free from external pressures in the decision-making process, the competent 'consenting' child is an ontological figure, brought into being and continuously shaped and re-shaped by the fast-evolving social and political landscape of disputed biological truths, the hegemony of queer theory, trans affirmative lobbying and trans activism, and the GIDS itself!

I prefer the less compromised view of the ethics of transgendering children which is put forward by Laidlaw. He asks, 'How can a child, adolescent or even parent provide genuine consent to such a treatment? How can the physician ethically administer gender affirming therapy, knowing that a significant number of patients will be irreversibly harmed?' He concludes:

> In our opinion, physicians need to start examining gender affirming therapy through the objective eye of the scientist-clinician rather than the ideological lens of the social activist. Far more children with gender dysphoria will ultimately be helped by such an approach. (Laidlaw et al, 2018)

Chapter One

Conclusion: ethics and the practice of transgendering children

The GIDS has helped invent the transgender child within the political psycho-dynamic forcefield of transactivism and transgender theory. Its alleged multi-factorial approach is in effect driven by a single theoretical construct: it has at its core the 'issue of identity' defined by transgender theory and lobbying. The GIDS subsumes the multifactorial physical, sociological and familial context within which a child identifies as transgender within an overall model of affirming transgenderism as an inherent unwilled phenomenon, a model which it is allegedly transphobic for sociologists, philosophers and psychologists to question.

The GIDS prides itself on steering a middle-course of 'watchful-waiting' to evaluate whether a child will persist or desist with identifying as transgender. In this chapter I have argued the alleged 'middle-course' is itself deeply politically positioned and ethically compromised—a phenomenon the clinic disavows when challenged. In resisting self-reflection on the psycho-dynamics of its own controversial practices, the GIDS has navigated ever closer to the terrain staked out by trans lobby groups and organisations, whose threat to the clinic it was Di Ceglie's initially stated intention to avoid in order to ensure the GIDS' continued survival.

Only time will tell whether the GIDS ship has navigated so near to the rocky island of trans ideology that the clinic will eventually bring about its own demise. The tide may be turning as more and more concerned people gain knowledge of the absence of evidence-based medicine, of the irreversible harms of hormone therapy, of dissent deep at the heart of the GIDS of whether the transgender child exists outside of political discourse.

What is urgent, as this book goes to press, is listening to those clinicians who have been attempting to verbalise their extreme disquiet. One clinician says: 'I keep thinking about all of the children, adolescents, families who are being harmed by the one-dimensional discussion and the attack on truth and on thinking, and on what we know about adolescent well-being…'. Another clinician replies:

> I absolutely agree with that. And I'm really angry at what's happening to these children. I'm angry with all the grown-ups, all the clever people, all the thoughtful people, who are officially letting this happen. What I've witnessed to me feels incredibly distressing and disturbing and like something that should be stopped. (Anonymous gender clinician, personal communication, 2018)

References

Brunskell-Evans, H. (2018). Gendered Mis-Intelligence: The Fabrication of 'The Transgender Child'. In H. Brunskell-Evans, and M. Moore (Eds.), *Transgender Children and Young People: Born in Your Own Body* (pp.41-63). Newcastle upon Tyne: Cambridge Scholars Publishing.

—. (2019). The Medico-Legal 'Making' Of 'The Transgender Child. In Special Issue of the *Medical Law Review* "Regulating Sexual Boundaries". Autumn: Open University Press.

Butler, G., De Graaf, N., Wren, B., and Carmichael, P. (2018). Assessment and support of children and adolescents with gender dysphoria. *Archives of Disease in Childhood*, *103*(7), 631-636.

Butler G., Wren B., and Carmichael P. (2019). Puberty blocking in gender dysphoria: suitable for all? *Archives of Disease in Childhood*. Published online January 17, 2019. doi: 10.1136/archdischild-2018-315984

Carmichael, P., Phillott, S., Dunsford, M., Taylor, A. and De Graaf, N. (2016). Gender Dysphoria in Younger Children: Support and Care in an Evolving Context. Presentation given at the World Association for Transgender Heath (WPATH) Conference, June 17-21. Retrieved from http://wpath2016.conferencespot.org/62620-wpathv2-1.3138789/t001-1.3140111/f009a-1.3140266/0706-000523-1.3140268

Department of Health (2008). *Medical Care for Gender Variant Children and Young People: Answering Families' Questions.* London: National Health Service.

Derman, S., Gamble, D., and Hakeem, A. (2002, July 15). The psychiatry of transsexuality [Letter to the editor]. *The Telegraph*.

De Vries, A.L., Steensma, T.D., Doreleijers, T.A., and Cohen-Kettenis, P.T. (2011). Puberty suppression in adolescents with gender identity disorder: a prospective follow-up study. *Journal of Sexual Medicine*, *8*(8), 2276-2283. doi: 10.1111/j.1743-6109.2010.01943.x

Di Ceglie, D. (2018). The use of metaphors in understanding atypical gender identity development and its psychosocial impact. *Journal of Child Psychotherapy, 44*(1), 5-28.
doi: 10.1080/0075417X.2018.1443151

Endocrine Society (2009). Endocrine Treatment of Transsexual Persons: An Endocrine Society Clinical Practice Guidelines. *The Journal of Clinical Endocrinology and Metabolism, 94*(9), 3132-3154.

—. (2017). Endocrine Treatment of Gender-Dysphoric/ Gender Incongruent Persons: An Endocrine Society Clinical Practice

Guideline. *The Journal of Clinical Endocrinology and Metabolism, 102*(11), 3869-3903

Fine, C. (2017). *Testosterone Rex: unmaking the myths of our gendered minds.* New York and London: Icon.

FOI (2018). Freedom of Information Disclosure Log, No: 16-17320. Retrieved from https://tavistockandportman.nhs.uk/documents/557/16-17320_Psychical_attacks_and_security_vk2fyj2.pdf

—. (2019). Early Pubertal Suppression in a Carefully Selected Group of Adolescents with Gender Identity Disorder, 4 November 2010, Research Ethics Committee number 10/H0713/79.

Gendered Intelligence (2012). *Trans Youth Sexual Health Booklet.* Retrieved from http://cdn0.genderedintelligence.co.uk/2012/11/17/17-14-04-GI-sexual-health-booklet.pdf

GIDS (2019). *Information for young people.* Retrieved from http://gids.nhs.uk/puberty-and-physical-intervention

GIRES (2018). *Gender Variance – for GPS.* Retrieved from http://www.gires.org.uk/e-learning/gender-variance-for-gps

House of Commons Women's Equality Commission (2016). *Transgender equality inquiry.* Retrieved from https://www.parliament.uk/business/committees/committees-a-z/commons-select/women-and-equalities-committee/inquiries/parliament-2015/transgender-equality/

Heneghan, C. and Jefferson, T. (2019). Gender Affirming Hormone in Children and Adolescents. *British Medical Journal Evidence Based Medicine Spotlight.* Retrieved from https://blogs.bmj.com/bmjebmspotlight/2019/02/25/gender-affirming-hormone-in-children-and-adolescents-evidence-review/

Laidlaw, M.K., Van Meter, Q.L., Hruz, P.W., Van Mol, A., and Malone, W.J. (2018). Letter to the Editor: "Endocrine Treatment of Gender-Dysphoric/Gender-Incongruent Persons: An Endocrine Society Clinical Practice Guideline". *The Journal of Clinical Endocrinology and Metabolism, 104*(3), 686-687. doi: https://doi.org/10.1210/jc.2018-01925

Memorandum of Understanding (2017). *Memorandum of Understanding on Conversion Therapy in the UK (version 2).* Retrieved from http:www.pinktherapy.com/portals/0/MoU2_Final.pdf

Moore, M. (2018). Standing Up for Boys and Girls. In H. Brunskell-Evans, and M. Moore (Eds.), *Transgender Children and Young People: Born in Your Own Body* (pp.218-232). Newcastle upon Tyne: Cambridge Scholars Publishing.

Matthews, S. (2018). The Body Factory: Twentieth Century Stories of Sex Change. In H. Brunskell-Evans, and M. Moore (Eds.), *Transgender Children and Young People: Born in Your Own Body* (pp.123-138). Newcastle upon Tyne: Cambridge Scholars Publishing.

NHS England (2015). *NHS Standard Contract For Gender Identity Development Service for Children and Adolescents.* Retrieved from https://www.england.nhs.uk/wp-content/uploads/2017/04/gender-development-service-children-adolescents.pdf

Richards C, Maxwell J., and McCune N, (2019). Use of puberty blockers for gender dysphoria: a momentous step in the dark. *Archives of Disease in Childhood*, January 17. doi: 10.1136/archdischild-2018-315881

Rippon, G. (2019). *The Gendered Brain.* London: Bodley Head.

Royal College of General Practitioners (2015). *Gender Variance E-Learning Module.* Retrieved from http://elearning.rcgp.org.uk/gendervariance

Tavistock and Portman NHS Trust. (2019). *Mission and values.* Retrieved from https://tavistockandportman.nhs.uk/about-us/who-we-are/mission-values

WPATH (2019). *Ethical Guidelines for Professionals.* Retrieved from https://www.wpath.org/about/ethics-and-standards

Wren, B. (2014). Thinking postmodern and practising in the enlightenment: Managing uncertainty in the treatment of children and adolescents. *Feminism & Psychology, 24*(2), 271-291.

—. (2018). The Complexity of Clinical Work with Gender Diverse Children and Adolescents. *University of Leicester Press Office.* Retrieved from https://www2.le.ac.uk/offices/press/press-releases/2018/january/experienced-psychologist-reflects-on-clinical-work-with-gender-variant-children-and-adolescents

Chapter Two

Britain's Experiment with Puberty Blockers

Michael Biggs

In 1994 a 16-year-old girl who wanted to be a boy, known to us as B, entered the Amsterdam Gender Clinic. She was unique for having her sexual development halted at the age of 13, because an adventurous paediatric endocrinologist had given her a Gonadotropin-Releasing Hormone agonist (GnRHa). Originally developed to treat prostate cancer, these drugs are also used to delay puberty when it develops abnormally early: in girls younger than 8, and in boys younger than 9. The innovation was to take the drugs to stop normal puberty altogether, in order to prevent the development of unwanted sexual characteristics—with the aim of administering cross-sex hormones in later adolescence. Dutch clinicians used B's case to create a new protocol for transgendering children, which enabled physical intervention at an age much earlier than the accepted age of consent (Cohen-Kettenis & Goozen, 1998).

The Dutch protocol promised to create a more passable simulacrum of the opposite sex than could be achieved by transition in adulthood. It was therefore embraced by trans-identified children and their parents, by older transgender activists, and by some clinicians specializing in gender dysphoria. The Gender Identity Development Service (GIDS), part of the Tavistock and Portman NHS Foundation Trust, treats children with gender dysphoria from England, Wales, and Northern Ireland. It launched an experimental study of 'puberty blockers'—the friendlier term for GnRHa when administered to children with gender dysphoria—in 2010. Fifty children were injected with triptorelin, for at least two years. This chapter describes the origins and conduct of this study and scrutinizes the evidence on its outcomes. It draws on information obtained by requests under the Freedom of Information Act to Tavistock Trust, to the NHS Health Research Authority, and to University College London (UCL). I will argue that the experimental study did not properly inform children and their parents of the risks of triptorelin. I will also demonstrate that the study's preliminary results were more negative than positive, and that the single article using data from the study is fatally flawed by a statistical fallacy. My conclusion is that GIDS and their collaborators at UCL have either ignored or suppressed negative evidence. Therefore, GIDS had no justification for introducing the Dutch protocol as general policy in 2014.

Origins

GnRHa drugs have never been licensed for treating children suffering from gender dysphoria. The particular drug used in Britain, as in the Netherlands, is triptorelin, which is licensed to treat advanced prostate cancer and sexual deviance in men; endometriosis and uterine fibroids in women (for no longer than six months); and precocious puberty in children (Electronic Medicines Compendium, 2019). Using GnRHa to treat gender dysphoria is 'a momentous step in the dark', for it is 'presumptuous to extrapolate observations from an intervention that suppresses pathologically premature puberty to one that suppresses normal puberty' (Richards et al., 2018). Therefore, the origins of Tavistock's experiment need some explanation.

The Dutch protocol became well known in Britain before the first scientific article was published. A television documentary showed trans-identified girls travelling to meet their peers in the Netherlands, who were taking GnRHa as young as 13 (Channel 4, 1996). This inspired Stephen Whittle—who led the transgender campaigning organization Press for Change—to argue for a legal right to access 'pubertal suppression'; doctors who failed to provide drugs could be vulnerable to litigation (Whittle & Downs, 2000; Wren, 2000, p.224). This argument was first advanced at a conference at Oxford in 1998, where the keynote speaker was the head of the Amsterdam Gender Clinic. There was little movement, however, over the next few years. Guidelines issued by the British Society for Paediatric Endocrinology and Diabetes (BSPED) in 2005 still insisted that children had to reach full sexual development (known as Tanner Stage 5)—around the age of 15—before being prescribed GnRHa drugs.

A crucial role was played by organizations that campaign for the transgendering of children: the Gender Identity Research and Education Society (GIRES) and Mermaids. GIRES organized a symposium in London in 2005 to develop 'guidelines for endocrinological intervention'. Additional funding came from Mermaids, two medical charities—the Nuffield Foundation and the King's Fund—and the Servite Sisters Charitable Trust Fund. This brought together the creators of the Dutch protocol, American clinicians like Norman Spack in Boston, and key British figures such as Domenico Di Ceglie, the Director of GIDS, and Polly Carmichael and Russell Viner, both at Great Ormond Street Hospital. (The latter two were to lead the 2010 experiment.) Some of the participants lobbied for the Dutch protocol. Veronica Sharp from Mermaids 'described users' and parents' views of the available treatments, and the anguish they may experience when hormone blocking is delayed' (GIRES, 2005). The symposium ended with an agreement to push for amendments to guidance from bodies like BSPED, and to conduct collaborative research between London, Amsterdam, and Boston. There was another meeting in Amsterdam the following year, but the collaborative research did not eventuate.

International developments did enable parents to circumvent the NHS. GIRES (2006) warned that 'those who can in any way afford to do so have to consider taking their children to the USA'. The first was Susie Green, who later became the chief executive of Mermaids. In 2007 she took her son Jackie, aged 12, to Boston, to purchase a prescription for GnRHa drugs from Spack; the drugs were supplied by an online Canadian pharmacy (Sloan, 2011). A presentation at Mermaids, presumably by Green, instructed parents in this medical tourism (Mermaids, 2007). Spack treated a further seven British children over the next few years (Glass, 2012).

By 2008, GIRES was more strident in criticizing British clinicians. One of its founders, Terry Reed, denounced them as 'transphobic':

> They are hoping that during puberty the natural hormones themselves will act on the brain to 'cure' these trans teenagers. What we do know is what happens if you don't offer hormone blockers. You are stuck with unwanted secondary sex characteristics in the long term and in the short term these teenagers end up suicidal. (Groskop, 2008)

Reed was clearly drawing on the experience of her own child, who had transitioned two decades before. This feature article in the *Guardian* signalled how the controversy was becoming newsworthy. GIRES objected to the fact that the Royal Society of Medicine's conference on gender dysphoria in adolescents had invited too few advocates for the Dutch protocol. The conference was noteworthy as the occasion for a rare public protest by transgender activists (Brown, 2018, p.311). The target was Kenneth Zucker from Toronto, a leading authority on gender dysphoria, who was denounced as a 'transphobic doctor who supports repression and torture of gender-variant children' (Kennedy, 2008). Activists were not the only critics. A medical ethicist at the University of

Manchester (who had attended the 2005 symposium) denounced Viner's caution about the risks of GnRHa, on the grounds that 'anything is better than life in an alien body' (Giordano, 2008, p.583). As the decade drew to a close, the demand for puberty blockers was irresistible.

Experiment

GIDS decided to frame the concession as research, undertaken in collaboration with scientists at UCL. Viner was the chief investigator; co-investigators included Carmichael, who had taken over as the Director of GIDS, and Di Ceglie, who had become the Director of Training, Development, and Research. The first proposal was rejected by the NHS Research Ethics Committee, on the grounds that it was not a proper randomized trial and therefore could not yield valid results (Young Minds, 2010). The second proposal—'Early pubertal suppression in a carefully selected group of adolescents with gender identity disorder' (Viner, 2010)—was no more rigorous. There was no random allocation of patients into control and treatment groups, and no double blinding of patients and medics. Nevertheless, the proposal was approved. It was not designed to maximize information on the effects of GnRHa. For example, children were asked to consent to complete questionnaires only until they were 16. If they had been asked to give consent for the researchers to access their medical records in perpetuity, then GIDS would have been able to analyze effects of the drugs over the long term. Although the proposal called this a 'study', I prefer the word 'experiment' to convey the fact that it was using a drug that was not licensed for this condition.

The research proposal provided a comprehensive review of the potential benefits and risks of GnRHa. 'It is not clear what the long term effects of early suppression may be on bone development, height, sex organ development, and body shape and their reversibility if treatment is stopped during pubertal development' (Viner, 2010, p.7). Viner spoke frankly in a later newspaper interview:

> If you suppress puberty for three years the bones do not get any stronger at a time when they should be, and we really don't know what suppressing puberty does to your brain development. We are dealing with unknowns. (Bracchi, 2012)

This caution echoed previous comments by Carmichael: 'the debate revolves around the reversibility of this intervention—physical and also psychological, in terms of the possible influence of sex hormones on brain and identity development' (Carmichael & Davidson, 2009, p.917).

When Tavistock Trust announced the study, however, it claimed that GnRHa treatment 'is deemed reversible' (Tavistock and Portman, 2011). More disturbing is the fact that the Patient Information Sheet provided to children when they gave consent also minimized or concealed the risks acknowledged in the research proposal.[1] Although the sheet ran to four pages, it omitted the fact that GnRHa drugs have never been certified as safe and effective for treating gender dysphoria. The words 'experiment' or 'trial' did not appear. Under 'the possible benefits of taking part' came this astonishing statement:

> If you decide to stop the hormone blockers early your physical development will return as usual in your biological gender [sic]. The hormone blockers will not harm your physical or psychological development.

This directly contradicted the chief investigator's own statements.

As for side effects, there was a vague warning that the drug 'could affect your memory, concentration and the way you feel'. The triptorelin formulations used by GIDS—Gonapeptyl® Depot and Decapeptyl® SR—carry detailed warnings of side effects. Depression is common, affecting between 1% and 10% of patients (Ferring Pharmaceuticals, 2016), and 'may be severe' (Ipsen, 2017). Other side effects affecting up to 10% of children treated for precocious puberty include 'pain in abdomen, pain bruising, redness and swelling at injection site, headache, hot flushes, weight gain, acne, hypersensitivity reactions' (Ipsen, 2017). None of these are mentioned in the Patient Information Sheet.

One further absence deserves emphasis. The 2005 Symposium had already noted the paradox that blocking a boy's puberty left him with stunted genitalia, which were then not sufficient to transform into a pseudo-vagina. 'Although there are surgical means to deal [with] this difficulty, the patient and her parents or guardians should be fully informed about its implications' (GIRES, 2005). The Patient Information Sheet failed to mention this.

All these omissions might be explained by the input of parents who saw GnRHa as an elixir that would enable their child to change sex. 'The wording ... was agreed with a number of families with whom the draft had been discussed' (Di Ceglie, 2019, p.149). Whatever the cause, GIDS and UCL gave children and parents incomplete and misleading information,

---

[1] Version 1.0, 4 November 2010, obtained from University College London under the Freedom of Information Act. One portion is reproduced by Di Ceglie (2019, p.149).

which contradicted the research proposal. Whether they could provide informed consent, in such circumstances, is open to serious question.

The course of the experiment can be gleaned from a conference presentation and a published abstract (Gunn et al., 2015a; Gunn et al., 2015b). From May 2010 to July 2014, 61 children were recruited, with a slight preponderance of boys.[2] GnRHa was administered to 50 of them; the others were too young, too thin, or had insufficient bone density. Under the Dutch protocol, children became eligible around age 12 (Tanner Stage 2 or 3). The age at which these subjects started the drug ranged from 10 to 16. None of the children started on the drug had ceased after two years.

Results

Before the final patient was enrolled, Carmichael announced success to the tabloid press. 'Now we've done the study and the results thus far have been positive we've decided to continue with it' (Manning & Adams, 2014). Her statement was misleading, at best. Six months earlier, Carmichael had already stated that she planned to continue the experiment indefinitely (Leake, 2013). Then the sole justification was the large number of parents demanding the drugs. At that point, only 23 children had taken triptorelin, so the trial was not even halfway through. These pronouncements make a mockery of Carmichael's earlier bromide: 'as professionals we need to be looking at the long term and making sure this treatment is safe' (Alleyne, 2011).

Where are these 'positive' results?[3] The current GIDS webpage on the evidence base for puberty blockers states that 'research evidence for the effectiveness of any particular treatment offered is still limited' (GIDS, 2019). There is no mention of its own experiment; it cites only research from the Netherlands. Di Ceglie stated last year that the 'project is ongoing and the results are yet to be published' (Di Ceglie, 2018, p.14).

Diligent searching does, however, uncover unpublished results. Most revealing is an appendix to Carmichael's report to Tavistock's Board of Directors (GIDS, 2015).[4] It tracks the first 44 children on triptorelin,

---

[2] Ethical permission was granted only in December 2010. Presumably, children who entered earlier waited for this permission to be granted before being injected with triptorelin. A cryptic graph implies that only 2 children were referred in 2010; 22 were referred in 2014 (Gunn et al., 2015a).
[3] I emailed the address listed on the webpage announcing the study (communications@tavi-port.nhs.uk) on 1 February 2018, inquiring after the results. There was no reply.
[4] My annotated version is available at

measuring changes after one year of the drug regime. The text is sometimes internally inconsistent and occasionally contradicts the tabulated figures, suggesting that the appendix was prepared in haste. But we can summarize those changes that were reported as statistically significant (p-value < .05). Only one change was positive: 'according to their parents, the young people experience less internalizing behavioural problems' (as measured by the Child Behavior Checklist). There were three negative changes. 'Natal girls showed a significant increase in behavioural and emotional problems', according to their parents (also from the Child Behavior Checklist, contradicting the only positive result). One dimension of the Health Related Quality of Life scale, completed by parents, 'showed a significant decrease in [the] Physical well-being of their child'. What is most disturbing is that, after a year on blockers, 'a significant increase was found in the first item 'I deliberately try to hurt or kill self'' (in the Youth Self Report questionnaire). Astonishingly, the increased risk of self-harm attracted no comment in Carmichael's report. Given that puberty blockers are prescribed to treat gender dysphoria, it is paradoxical that 'the suppression of puberty does not impact positively on the experience of gender dysphoria' (measured by the Body Image Scale). When differentiated by sex, the impact was positive for boys on one aspect of body image, but negative for girls on two aspects.

These preliminary results (44 children after one year on triptorelin) also appear in an abstract for the World Professional Association for Transgender Health:

> For the children who commenced the blocker, feeling happier and more confident with their gender identity was a dominant theme that emerged during the semi-structured interviews at 6 months. However, the quantitative outcomes for these children at 1 years time suggest that they also continue to report an *increase in internalising problems and body dissatisfaction* [my emphasis], especially natal girls. (Carmichael et al., 2016)

These findings pertain to 44 out of 50 of the children in the experiment. It is exceedingly unlikely that they would be altered by the inclusion of the last 6 subjects. Moreover, children and parents had a clear bias towards reporting favourable outcomes; after all, they had enrolled in the experiment because they viewed GnRHa drugs as beneficial. This positive bias increases the probative value of negative evidence. Why were these negative results never published?

---

http://users.ox.ac.uk/~sfos0060/Annotated_GIDS_results.pdf.

One article on the outcome of puberty blockers, coauthored by Carmichael, apparently includes some data from the experiment (Costa et al., 2015). The article discusses 101 children given GnRHa drugs at GIDS, starting at ages ranging from 13 to 17. Given the date of publication, most or all of those children who started at ages 13 and 14 (and perhaps 15?) must have been part of the 2010–14 experiment. But the age range also indicates the exclusion of some of the experiment's children: those who commenced GnRHa from ages 10 to 12. Excluding some subjects without justification is poor practice and raises the suspicion of cherry picking. Nevertheless, we could consider this article as having some bearing on the 2010–14 experiment.

The abstract proclaims that 'adolescents receiving also puberty suppression had significantly better psychosocial functioning after 12 months of GnRHa ... compared with when they had received only psychological support' (Costa et al., 2015, p.2206). The article is treated in the literature as providing evidence in favour of puberty blockers (e.g. Butler et al., 2018; Heneghan & Jefferson, 2019). But the abstract is misleading: the analysis actually *failed to detect any difference* between children who were given blockers and those who were not. To understand this, we need to scrutinize the article in detail. (Statistically minded readers will recognize the fallacy described by Gelman & Stern, 2006.)

The analysis starts with 201 adolescents diagnosed with gender dysphoria. The children were divided into two groups: those deemed eligible for puberty blockers immediately, and those who needed more time due to 'comorbid psychiatric problems and/or psychological difficulties'. This second group did not receive any physical intervention during the time of analysis, and so serves as a comparison group. Both groups received psychological support. The article chooses one outcome: psychosocial functioning as measured by the Children's Global Assessment Scale (CGAS). This scale was administered at the outset, and then after six, twelve, and eighteen months. It is suspicious that the article omits all the outcomes that were negative in the preliminary results of the 2010–14 experiment: the Child Behavior Checklist, the Youth Self Report Questionnaire, the Health Related Quality of Life scale, and the Body Image Scale.

The authors graph the CGAS results, but without confidence intervals—which indicate the extent of random statistical variation or noise. (The graph is redrawn with confidence intervals in Biggs, 2019.) The smaller the sample, the greater this noise. These samples shrank over time: after eighteen months, the group getting drugs numbered only 35, and the comparison group 36. The article does not explain why two thirds of the subjects disappeared. Presumably they did not stop the medication, because all the children in the 2010–14 experiment continued the drug regime for two years (Gunn et al., 2015b).

The group given puberty blockers from six months onwards showed improvement at eighteen months: the average CGAS score had increased from 61 to 67. This improvement is statistically significant, and it is the one that the authors chose to highlight. However, these children also received psychological support, and so attributing this improvement to medical intervention is unjustified. The crucial comparison is between the group receiving blockers and the comparison group. The latter's average CGAS score after eighteen months was lower, 63 compared to 67. This is hardly surprising because the comparison group was composed of children with more serious psychological problems. Anyway, this difference is *not statistically significant*: a two-tailed *t*-test for the difference between group means yields a *p*-value of .14, far beyond the conventional .05 threshold. In other words, the samples were so small, and there was such wide variation in scores within each group, that we can draw no conclusions. There is no evidence that puberty blockers improve psychosocial functioning. No wonder that GIDS' own webpage on the evidence for medical intervention does not cite this article (GIDS, 2019).

The failure to fully publish the results of the experiment—for all 50 children given triptorelin, on all the outcomes that the study measured—suggests that it was a pretext to administer unlicensed drugs rather than an attempt to gain scientific knowledge.

Consequences

The failure to publish comprehensive results would be serious even if the unlicensed use of triptorelin had been confined to the 50 experimental subjects. However, the Director of GIDS took part in a BBC television documentary—aimed at children aged 6 to 12—broadcast in November 2014. It followed a trans-identified girl aged 13, Leo, who was one of the experimental subjects. Carmichael appears talking to Leo in reassuring tones:

> The blocker is an injection that someone has every month which pauses the body and stops it from carrying on to grow up into a man or a woman. … And the good thing about it is, if you stop the injections, it's like pressing a start button and the body just carries on developing as it would if you hadn't taken the injection. (BBC, 2014)

To emphasize this point for the juvenile audience, the film superimposes a pause button on the screen. Viner's earlier comment bears repeating: 'If you suppress puberty for three years the bones do not get any stronger at a time when they should be, and we really don't know what suppressing puberty does to your brain development' (Bracchi, 2012). Needless to say, Carmichael does not tell Leo that children in the experiment were more likely to self-harm after a year on triptorelin, nor that girls experienced greater dysphoria.

Tavistock Trust then embraced the Dutch protocol with enthusiasm. Three years later, GIDS (and its satellite operation in Leeds) had prescribed puberty blockers for a total of 800 adolescents under 18, including 230 children under 14 (Manning, 2017). By 2018, new prescriptions were running at 300 per year (BBC News, 2018). Freedom of Information requests have failed to elicit more recent figures because GIDS does not collate basic data on this experimental treatment—nor does the University College London Hospitals NHS Foundation Trust, which provides its endocrinology services. Apparently 'work is currently in progress to manually enter all hormone blocker prescription data onto a database, pending future meetings with UCLH and LGI [Leeds General Infirmary] to ascertain who is collecting this info and how it is to be reported.'[5]

The abstract describing the baseline characteristics of the children in the experiment concluded: 'Assessment of growth, bone health and psychological outcomes will be important to assess the medium and *long-term safety and effectiveness* of early intervention' (Gunn et al., 2015b, A198, my emphasis). This aspiration was never implemented. GIDS recently acknowledged that it loses track of its patients after they turn 18, blaming 'the frequent change in nominal and legal identity, including NHS number in those referred on to adult services'—'to date they have not been able to be followed up' (Butler et al., 2018, p.635).[6] By contrast, the Amsterdam clinic carefully tracks its patients over time. The pioneer, B, has been followed to the age of 35. He did not regret transition, but scored high on the measure for depression. Owing to 'shame about his genital appearance and his feelings of inadequacy in sexual matters', he could not sustain a romantic relationship (Cohen-Kettenis et al., 2011, p.845). To the clinicians, however, this case exemplifies the success of the Dutch protocol.

---

[5] Internal Review of Freedom of Information request (18-19312) submitted by Susan Matthews to Tavistock Trust, 24 February 2019.

[6] Transgender activists successfully lobbied the NHS to provide new numbers to patients as well as to change the sex on their medical records (Birch, 2014).

Conclusion

GIDS and UCL launched an experiment in 2010 to use GnRHa drugs to stop puberty. The impetus for this unlicensed treatment came from children and parents, along with transgender activists and some clinicians, who seized on the notion that blocking puberty was akin to alchemy—it would enable a child to change sex, as long as he or she started young. Given the unrelenting pressure from Mermaids and GIRES, supported by the climate of opinion among the *Guardian*-reading classes, Tavistock arguably had to concede to the demand for GnRHa below the age of 16. From the outset, however, the experiment was flawed. The Patient Information Sheet understated the risks of this unlicensed treatment, despite those risks being acknowledged explicitly in the research proposal. Worse was to come. Before the experiment had run its course, Carmichael claimed 'that the results thus far have been positive' in order to justify what must have been a premeditated decision to incorporate the Dutch protocol into the policy of GIDS. She even appeared on children's television to promote GnRHa drugs.

In fact, the experiment showed predominantly negative outcomes (GIDS, 2015). After a year on triptorelin, children reported greater self-harm; girls also experienced more behavioural and emotional problems and expressed greater dissatisfaction with their body—so drugs exacerbated gender dysphoria. The fact that these outcomes have never been published is a serious indictment of Carmichael, Viner (now President of the Royal College of Paediatrics and Child Health), Di Ceglie, and the other scientists who proposed the research.[7] The failure can be highlighted by comparing another use of triptorelin: the treatment of hypersexuality in men, for which it is licensed. The chemical castration of seven dangerous sex offenders in Broadmoor Hospital resulted in a report spanning two pages, which detailed the adverse side effects experienced by three patients (Ho et al., 2012). The use of triptorelin on 50 adolescents—off license—produced only a half-page published abstract (Gunn et al., 2015b). Some of the experimental subjects were apparently included with older adolescents from GIDS in one published analysis (Costa et al., 2015). It examines a single outcome measure—notably not one of the measures that yielded negative effects in the preliminary results. This article misrepresents its finding. Properly analyzed, it shows no evidence for the effectiveness of the drugs: there was no statistically

---

[7] Names were redacted in the copy obtained from the Health Research Authority.

significant difference in psychosocial functioning between the group given triptorelin and the comparison group given only psychological support.

My critique has evaluated Tavistock's experiment in accord with its own aims, as laid out in the 2010 research proposal. For reasons of space, this chapter has not discussed three additional problems attending the use of GnRHa drugs to block puberty. The Dutch protocol was originally touted as a diagnostic aid as well as a treatment; it would give the child time to ponder her or his gender identity (Cohen-Kettenis, 1998). In fact, however, children given GnRHa drugs almost invariably progress to cross-sex hormones. The 2010–14 experiment was typical insofar as none of the children stopped the drug regime within two years. (GIDS never revealed the proportion who went on to cross-sex hormones.) Before the introduction of puberty blockers, around four fifths of young children with gender dysphoria would grow out of it naturally, typically becoming gay, lesbian, or bisexual adults (e.g. Zucker, 2018). Using GnRHa drugs to block puberty does not mean pressing a pause button, as Carmichael asserted—it is more like pressing fast forward into cross-sex hormones and ultimately surgery.

The second problem is obvious. Blocking puberty effectively destroys the individual's ability to have children. If the adolescent stops taking GnRHa, fertility should recover, but as we have seen, stopping is exceptional. The third problem is rarely admitted. Blocking puberty impedes the development of sexual functioning; some children given GnRHa drugs never develop the capacity for orgasm (Jontry, 2018). There is a strong taboo against mentioning this. The word did not appear in the proposal for the 2010–14 experiment, and never appears on the GIDS website. When the endocrinologist at GIDS, Gary Butler, was asked about the effect of GnRHa on the ability to orgasm, he refused to answer.[8]

When Tavistock Trust was presented with my critique (Biggs, 2019), it failed to address any of the specific charges. It claimed that 'GIDS is actively contributing to the evidence base to inform the best way to support gender-diverse young people' (Tominey & Walsh, 2019). Yet GIDS' own webpage on the evidence base for GnRHa drugs does not mention its own experiment, nor does it cite its own article. The Trust also boasted of winning '£1.3 million to conduct research with the University College London and the Universities of Liverpool and Cambridge into the long-term outcomes for young people who use the service.' There can be no confidence in the ability of GIDS to track its own patients over the long

---

[8] The question was posed by Susan Matthews after Butler's talk to the European Society for Paediatric Endocrinology's symposium on the Science of Gender, London, 19 October 2018.

term—recall that it cannot keep a tally of the number of children on triptorelin—let alone to publish the results. Tavistock Trust has failed not just the scientific community, but more importantly the children in its care.

Acknowledgements

My thanks are due to Stephanie Davies-Arai, Susan Matthews, and Elin Lewis for sharing their extensive knowledge, and to Sherena Kranat for steadfast support.

References

Alleyne, R. (2011, 15 April). Puberty blocker for children considering sex change. *Daily Telegraph*. Retrieved from https://www.telegraph.co.uk.
Biggs, M. (2019). Tavistock's experimentation with puberty blockers: Scrutinizing the evidence. Retrieved from http://www.transgendertrend.com/tavistock-experiment-puberty-blockers/
Birch, R. (2014). Q&A: Recording gender in medical records. Retrieved from http://www.pulsetoday.co.uk/your-practice/regulation/qa-recording-gender-in-medical-records/20008359.article
Bracchi, P. (2012, 25 February). Mixed-up five-year-olds and the alarming growth of the gender identity industry. *Daily Mail*. Retrieved from https://www.dailymail.co.uk.
Brown, S. (2018). The activist new wave. In Christine Burns (Ed.), *Trans Britain: Our journey from the shadows* (pp.304–16). Cornerstone.
Butler, G., de Graaf, N. Wren, B. & Carmichael, P. (2018). Assessment and support of children and adolescents with gender dysphoria. *Archives of Diseases of Childhood, 103*(7), 631–6.
BBC (2014). *My life: I am Leo* [television programme]. UK: Nine Lives Media.
BBC News (2018, July 2). Transgender children: Buying time by delaying puberty. *BBC News*. Retrieved from https://www.bbc.co.uk/
Carmichael, P. & Davidson, S. (2009). A gender identity development service. *The Psychologist, 22*(11), 916–7.
Carmichael, P., Phillott, S. Dunsford, M. Taylor, A., & de Graaf, N. (2016). Gender dysphoria in younger children: Support and care in an evolving context. World Professional Association for Transgender Health, 24th Scientific Symposium. Retrieved from http://wpath2016.conferencespot.org/62620-wpathv2-1.3138789/t001-1.3140111/f009a-1.3140266/0706-000523-1.3140268

Channel 4 (1996). *The decision: The wrong body* [television programme]. UK: Windfall Films.
Cohen-Kettenis, P. T. & van Goozen, S. H. M. (1998). Pubertal delay as an aid in diagnosis and treatment of a transsexual adolescent. *European Child and Adolescent Psychiatry*, 7(4), 246–48.
Cohen-Kettenis, P. T., Schagen, S. E. E., Steensma, T. D., de Vries, A. L. C., & Delemarre-van de Waal, H. A. (2011). Puberty suppression in a gender-dysphoric adolescent: A 22-year follow-up. *Archives of Sexual Behavior*, 40(4), 843–47.
Costa, R., Dunsford, M., Skagerberg, E., Holt, V., Carmichael, P., & Colizzi, M. (2015). Psychological support, puberty suppression, and psychosocial functioning in adolescents with gender dysphoria. *Journal of Sexual Medicine*, 12(11), 2206–14.
Di Ceglie, D. (2018). The use of metaphors in understanding atypical gender identity development and its psychosocial impact. *Journal of Child Psychology*, 44(1), 5–28.
—. (2019). Autonomy and decision making in children and adolescents with gender dysphoria. In M. Shaw & S. Bailey (Eds.), *Justice for Children and Families: A Developmental Perspective* (pp.145–53). Cambridge: Cambridge University Press.
Delemarre-van de Waal, H. A. & Cohen-Kettenis, P. T. (2006). Clinical management of gender identity disorder in adolescents: A protocol on psychological and paediatric endocrinology aspects. *European Journal of Endocrinology*, 155(suppl), S131–37.
Electronic Medicines Compendium (2019). Triptorelin. Retrieved from http://www.medicines.org.uk/emc/search?q=triptorelin#
Ferring Pharmaceuticals Ltd (2016). *Package leaflet: ... Gonapeptyl® Depot 3.75mg*. Retrieved from http://www.medicines.org.uk/emc/product/2229/pil
Gelman, A. & Stern, H. (2006). The difference between 'significant' and 'not significant' is not itself statistically significant. *American Statistician*, 60(4), 328–31.
GIDS (2015). Preliminary results from the early intervention research. In Tavistock and Portman Foundation NHS Trust, *Board of Directors part one: Agenda and papers ... 23rd June 2015* (pp.50–55).
—. (2019). *Evidence base*. Retrieved from http://gids.nhs.uk/evidence-base
GIRES (2005). Consensus report on symposium in May 2005. Retrieved from http://www.gires.org.uk/consensus-report-on-symposium-in-may-2005/

—. (2006). GIRES final report to the Nuffeld Foundation. Retrieved from http://www.gires.org.uk/gires-final-report-to-the-nuffield-foundation/

Glass, K. (2012, 22 January). A boy's own story. *The Times.* Retrieved from https://www.thetimes.co.uk.

Groskop, V. (2008, 14 August). "My body is wrong". *Guardian.* Retrieved from https://www.theguardian.com.

Gunn, H.M., Goedhart, C. Butler, G., Khadr, S.N., Carmichael, P.A., & Viner, R.M. (2015a). Gender dysphoria: Baseline characteristics of a UK cohort beginning early intervention. Presented to the Youth Health Conference, Australia. Retrieved from http://repository.tavistockandportman.ac.uk/1156/

—. (2015b). Early medical treatment of gender dysphoria: baseline characteristics of a UK cohort beginning early intervention. *Archives of Disease in Childhood, 100*(supp.3), A198.

Heneghan, C. & Jefferson, T. (2019). Gender-affirming hormone in children and adolescents. *British Medical Journal Evidence Based Medicine Spotlight.* Retrieved from http://blogs.bmj.com/bmjebmspotlight/2019/02/25/gender-affirming-hormone-in-children-and-adolescents-evidence-review/

Ho, D. K., Kottalgi, G., Ross, C. C., Romero-Ulceray, J. and Das, M. (2012). Treatment with triptorelin in mentally disordered sex offenders: Experience from a maximum-security hospital. *Journal of Clinical Psychopharmacology, 32*(5), 739–40.

Ipsen Ltd (2017). *Package leaflet: ... Decapeptyl® SR 11.25 mg.* Retrieved from http://www.medicines.org.uk/emc/product/780/pil

Jontry, B. (2018, July 8). Does prepubertal medical transition impact adult sexual function? *4thWaveNow.* Retrieved from http://4thwavenow.com/

Kennedy, N. (2008). Protest against transphobic psychologist Kenneth Zucker in London, 1st October. Retrieved from http://www.indymedia.org.uk/en/regions/london/2008/09/409405.html

Leake, J. (2013, 17 November). NHS helps children choose their sex. *Sunday Times.* Retrieved from https://www.thetimes.co.uk.

Manning, S. (2017, 30 July). How 800 children as young as 10 have been given sex change drugs. *Mail on Sunday.* Retrieved from https://www.dailymail.co.uk.

Manning, S. & Adams, S. (2014, 17 May). NHS to give sex change drugs to nine-year-olds. *Mail on Sunday.* Retrieved from https://www.dailymail.co.uk.

Mermaids (2007). Obtaining help from the Children's Hospital Boston. Presentation to the Mermaids annual meeting. Retrieved from http://www.gires.org.uk/wp-content/uploads/2014/08/mermaids-presentation.ppt

Richards, C., Maxwell, J., & McCune, N. (2019). Use of puberty blockers for gender dysphoria: A momentous step in the dark. *Archives of Disease in Childhood.* http://dx.doi.org/10.1136/archdischild-2018-315881

Sloan, J. (2011, 19 October). I had sex swap op on my 16th birthday. *Sun.* Retrieved from https://www.thesun.co.uk.

Tavistock and Portman NHS Foundation Trust (2011). Gender Identity Development Service conducts new research. Retrieved from http://tavistockandportman.nhs.uk/about-us/news/stories/gender-identity-development-service-conducts-new-research/

Tominey, C. & Walsh, J. (2019, 7 March). NHS transgender clinic accused of covering up negative impacts of puberty blockers on children by Oxford professor. *The Telegraph.* Retrieved from https://www.telegraph.co.uk

Viner, R. (2010). Early pubertal suppression in a carefully selected group of adolescents with gender identity disorder. Proposal submitted to Central London REC 2, 5 November 2010, Research Ethics Committee number 10/H0713/79

Whittle, S. & Downs, C. (2000). Seeking a gendered adolescence: Legal and ethical problems of puberty suppression among adolescents with gender dysphoria. In Eric A. Heinze (Ed.), *Of Innocence and Autonomy: Children, Sex and Human Rights* (pp.195–208). Aldershot: Ashgate.

Young Minds (2010). Tavistock bid for controversial drugs trial rejected by NHS on technicality. *Young Minds, 108*, 4.

Zucker, K. J. (2018). The myth of persistence. *International Journal of Transgenderism 19*(2), 231–45.

Chapter Three

Transgender Children: The Making of a Modern Hysteria

Lisa Marchiano

As a new clinician, I came across the following case report from the early days of psychoanalysis:

> At eighteen, her condition had got so bad that she really did nothing else than alternate between deep depressions and fits of laughing, crying, and screaming. She could no longer look anyone in the face, kept her head bowed, and when anybody touched her stuck her tongue out with every sign of loathing. (Jung, 1961, p.21)

This young woman was exhibiting symptoms in 1904 that today would likely earn her a diagnosis of schizophrenia. Her prognosis would be very poor. I was surprised to learn that this young woman made a full recovery within months, and went on to earn a medical degree from the University of Zurich. She herself became an important psychoanalyst, collaborating with three giants of 20th century psychology—Jung, Freud, and Piaget. How could someone so ill have gone on to have made such extraordinary contributions?

I could not make sense of this case because I did not know then what I know now—that virtually all psychiatric disorders are to some extent artifacts of the time and place in which they occur. The instinctual energies that arise from the unconscious are protean in form. They make their way into consciousness via vague sensations, mysterious moods, strange impulses, or fleeting images. When we suffer from underlying psychosocial vulnerabilities, disruption, trauma, or interpersonal anguish, our unconscious looks for culturally sanctioned garb in which to clothe our distress. Symptoms gain cultural currency through a complex and largely unconscious negotiation between the medical establishment, activists and advocates, media, and the patients themselves. Once these symptom templates have been codified and validated, they can be found by those unconsciously seeking to express wordless distress, and a feedback loop begins, further reifying the condition.

In this chapter, I aim to place the concept of the transgender child into the context of what literary critic Elaine Showalter calls 'hysterical epidemics' (1997). I explore how the necessary ingredients for such an epidemic have recently coalesced to create ideal conditions for the growth of a new hysteria—that of the transgender child.

Hysteria

The patient described in the first paragraph was Sabina Spielrein, a brilliant young Russian woman who was sent to the Burgholzli Psychiatric Clinic in Zurich at eighteen years of age. Her symptoms presented no mystery to Carl Jung, who wrote the cited case report. Spielrein was exhibiting the classic symptoms of hysteria. She was a textbook case of what was, at that time, the most common psychiatric diagnosis in Europe. Hysteria had long been a catch all diagnosis, but in the 19th century, it was codified, categorized—and exalted—by doctors like Jean-Martin Charcot. Its heightened profile corresponded with a significant increase in the number of women given the diagnosis who all presented the dramatic symptoms that Charcot had popularized (Shorter, 1993). By 1904, when Sabina Spielrein arrived in Zurich, hysteria was still in its heyday.

In treating Spielrein, the young Dr. Jung drew on the theories of his famous older mentor Sigmund Freud, who believed that hysterical symptoms were the result of intrapsychic conflict due to the repression of sexual fantasies. However, once Jung became more intimate with the inner world of his patient, Jung realized that Spielrein's experience could not fit so easily into the categories delimited by Freud or Charcot. In the dreams and symptoms of Spielrein and his other patients, Jung saw evidence of the autonomous workings of the unconscious, which spontaneously offered up images of synthesis, healing, and wholeness. His work with Spielrein contributed to his understanding of the psyche as a self-regulating system, wherein symptoms are metaphorical communications from the unconscious aimed at bringing us into balance.

By choosing to use the word 'hysteria' to describe the recent trend of transgender children and teens, I am locating this phenomenon within a tradition of clinicians and thinkers who have carefully documented the role of culture in shaping and spreading psychological symptoms. These writers have noted the psyche's tendency to express itself through the language made available to it by current cultural conceptualizations of distress. All of them speak to the dangers of well-intentioned attention to these kinds of contagions. Privileging the symptoms only makes them more entrenched and likely to spread, claiming more victims (Watters, 2011).

Is it wrong to put childhood gender dysphoria in the same category as 19th century hysterics, late 20th century anorexics, and victims of satanic ritual abuse? I do not believe that it is. While I find it plausible that acute gender dysphoria—unexplained, mysterious, and extremely painful—has always been with humanity, I suspect the number of genuine sufferers has been very small. Like other disturbances in the mind-body relationship such as anorexia or body dysmorphia, gender dysphoria appears to have a significant genetic component. Shorter has theorized that some people may be 'genetically predisposed to acquire some kind of disturbance of the mind-body relationship' (1993, p.193). That such a predisposition might be a common root of the three conditions is an intriguing possibility supported by the co-occurrence of eating disorders and gender dysphoria (Feder et al, 2017). Yet, while there is likely a biological component to gender dysphoria, the cultural narrative surrounding it will in large part determine how it is experienced by sufferers.

The protean nature of the unconscious

The epidemics of psychogenic diseases that have occurred in the recent past—including eating disorders, multiple personality disorder, satanic ritual abuse, and transgenderism—have in common a disavowal of the unconscious. When we do not consider the unconscious, we have no choice but to understand symptoms within the impoverished vocabulary of their concrete presentation. Instead of listening to the subtle communications that come to us from the body that can alert us to goings on in our inner world, we understand them as a problem with the body—a body that, in the case of transgenderism, then requires medical intervention. Medicalizing a psychological problem necessitates a medicalized treatment pathway, closing down opportunities for multiple possible meanings.

Jung came to be the great articulator of the fluid and mercurial nature of the unconscious. The powerful energies that comprise our unconscious never lose their potency, though they may clothe themselves in different forms depending on the age. In a remarkably prescient passage, Jung points us to the dangers that come from discounting these invisible forces.

> We think we can congratulate ourselves on having already reached such a pinnacle of clarity, imagining that we have left all these phantasmal gods far behind. But what we have left behind are only verbal spectres, not the

psychic facts that were responsible for the birth of the gods. We are still as much possessed by autonomous psychic contents as if they were Olympians. Today they are called phobias, obsessions, and so forth; in a word, neurotic symptoms. The gods have become diseases; Zeus no longer rules Olympus but rather the solar plexus, and produces curious specimens for the doctor's consulting room, or disorders the brains of politicians and journalists who unwittingly let loose psychic epidemics on the world. (1967, p.37)

The ancients experienced themselves as being under the constant influence of god-like forces. Today, the 'gods' speak to us through symptoms such as disturbances in the mind-body connection. Whereas these energies took the form of gods in the ancient world, or witchcraft and spirit possession in pre-modern Europe, today we speak of chemical imbalances in the brain, or say that we are born in the wrong body.

Jung astutely intuited that by undervaluing these energies, we become vulnerable to psychic epidemics. We do not understand the strength of the thing that has gripped us, and in our arrogance, assume that our narrow ego attitude must be the correct one. This helps explain why even seemingly fantastical ideas—that Satanists have infiltrated daycares, or that one can be born in the wrong body—can take hold of the collective.

Does it matter what name we give our suffering? Spielrein had available to her the template of 'hysteria', which seemingly gave her access to a treatment that was successful. Yet not all templates are benign. Contagions that lead us into damaging behaviors that become compulsive can be dangerous. Anorexia—which was exceedingly rare in the 20th century before it was 'popularized' in the 1970s—has the highest mortality rate of any psychiatric disorder (Arcelus et al, 2011). Moreover, where the treatment for a psychogenic disease involves invasive measures that permanently alter one's endocrine system, destroy one's fertility, or remove healthy tissue, we ought to be circumspect before taking the symptoms at face value. Instead, we would do well to be curious about what the psyche might be trying to tell us. This will require us to cultivate a symbolic rather than a literal attitude toward our suffering, as I have written about in a previous contribution to an earlier book (Marchiano, 2018).

Hysteria as communication

'The psyche is never silent', says Jungian analyst Jim Hollis. 'It is always speaking. It speaks through the body, dreams, symptoms, and intuitions. It keeps knocking on the door' (Selmakuvar, 2014). When a young person suffers from gender dysphoria, we must be willing to entertain the possibility that her symptoms are not due entirely to external factors such as genetics. Rather, we should consider that the symptoms she is manifesting are those carefully chosen by a psyche attempting to formulate something for which it has no words.

Some of these communications may be interpersonal, telegraphing to parents a need for greater independence amidst a terror of separation. Other messages might be meant for peers, communicating a need to belong, as well as to stand out. Finally, hysterias including gender dysphoria can make cultural statements, giving voice to tensions inherent at a particular moment in collective life.

Showalter writes that 'mass psychogenic disorders are metasymbols of the deep structures of our culture' (1997, p.8). Just as Jung explored what the spontaneous appearances of UFOs said about the psychic state of the collective in the aftermath of World War II (Jung & Adler, 1970), we might wonder what a culture's predominant mode of psychogenic illness communicates about its deepest desires, wounds, and fears. Many clinicians have noted the depth of metaphor and meaning communicated by the symptoms of anorexia and bulimia. And what of the symbolic nature of gender dysphoria?

At its core, gender dysphoria speaks to a profound loss of connection with our embodied, instinctual selves. The assumptions of transgender treatment require us to accept the body as a 'meat golem.' The animating principle has withdrawn from the body, and in the latter, we find only suffering and disgust. The living body has become mere clay which must be surgically and chemically altered to bring it into line with its master, the mind. The hypertrophied mind, cut off from its own instinctual, embodied base, disavows its connection to nature, instinct—and the unconscious. Paradoxically, the symptoms of gender dysphoria may in part be an attempt of the unconscious to reassert itself and signal that something is amiss and needs our attention.

Symptom pool

In seeking to give voice to anguish, the unconscious chooses from available symptoms that have achieved cultural validity. Like an improv actor who must select props and a costume from a jumble of items, the unconscious reaches into the cultural symptom pool to find ways to outfit itself to best convey its pain. Shorter coined the term 'symptom pool' to refer to those presentations which are available to members of a given culture in a given time and place as a legitimate and consensually sanctioned way of expressing distress. The symptom pool is 'the culture's collective memory of how to behave when ill' (1993, p.2). In medieval Europe, a common malady was the belief that one was made of glass and needed to be carefully swaddled to avoid breakage (Inglis-Arkell, 2014). In Indonesia, Malaysia, and New Guinea, men who have been dealt a blow or are under stress might evidence 'amok', a pattern of destructive and explosive rage (Gordon, 2000). Our early 21st century symptom pool includes the notion that children can suffer extreme distress as a result of being born in the wrong body.

Although there have likely always been children who were distraught about their gender, the notion of the transgender child entered the symptom pool only very recently. Gender identity disorder of childhood first appeared in the *Diagnostic and Statistical Manual of Mental Disorder III*, published in 1980, but for many years, transgender children were not part of the popular imagination. By 2015, however, the notion of the transgender child had become an uncontroversial idea.

Why do some symptoms gain cultural currency and enter the symptom pool and not others? We can understand hysterias better by exploring what they have in common. In examining medically shaped contagions such as 19th century hysteria and 20th century anorexia, we see that we need as a prerequisite a disease or condition with vague symptoms that cannot be objectively verified. Then, for this condition to take root and spread as a contagion, Showalter identifies three main ingredients: 'physician enthusiasts and theorists; unhappy, vulnerable patients; and supportive cultural environments' (1997, p.17). One can see the same components at work in hysterias, whether one looks at the Tanzanian laughing sickness, bulimia, or alien abduction. To round out our exploration of childhood gender dysphoria as a modern hysteria, I will contrast it in the following sections with late 19th century hysteria, as codified and made famous by the eminent French neurologist, Jean-Martin Charcot.

Subjective symptoms

Even in the 19th century, hysteria was regarded by some physicians with suspicion, and viewed as a wastebasket diagnosis (Scull, 2011). In Paris, Jean-Martin Charcot at first worked hard to identify via autopsy anatomical evidence of lesions that could explain hysterical symptoms. He found none (Shorter, 1993). Hysterical symptoms were vague and varied so that nearly any unusual presentation that was otherwise unexplainable might be ascribed to hysteria. Symptoms were either behavioral or based upon patients' reports. These might include headaches, partial paralysis, irritability, blindness in one eye, loss of sensation on one side of the body, fits of crying or screaming, an inability to distinguish colors, nervousness, insomnia, ovarian pain, fatigue, excessive yawning, and so on (Shorter, 1993). Charcot claimed to be able to treat hysteria with ovarian manipulation, hypnosis, and metallotherapy. Of course, all the behavioral symptoms and their cures could be effected by physician suggestion.

Psychogenic conditions manifest through subjective symptoms that have no plausible organic explanation (Weir, 2005). To offer a compelling choice for the unconscious, psychogenic illnesses must consist of symptoms that cannot be refuted by any objective measure, otherwise, they will not be credible. The symptoms of gender dysphoria are subjective and unfalsifiable. In a teenager, the diagnosis is made entirely on the basis of the young person's self-report. For example, a teen natal female might state that she cannot stand to get undressed in a room with a mirror because she hates the sight of her female body. She may say she knows that she really is a boy. She may not be able to get out of bed to go to school when she has her period. She may feel depressed, become withdrawn, or express a desire to self-harm. She might report to a therapist that she has 'always felt this way', and recall not liking dresses as a child. These symptoms and behaviors might concern us, but we could not identify a simple cause for these feelings. There is no identifiable pathogen, no blood test, and no x-ray that will definitively diagnose this condition, the treatment of which often involves invasive procedures. Even if we were to label these feelings 'gender dysphoria', such a label would not tell us anything about the underlying reasons for these feelings, or their true nature. This makes these feelings and behaviors compelling possibilities for the symptom pool, since they cannot be objectively contradicted.

Physician enthusiasts and theorists

Throughout recent history, physicians and therapists have played a leading role in validating symptoms and behaviors as legitimate components of the symptom pool. Jean-Martin Charcot (1825-1893) was the unwitting evangelist of hysteria in the 19th century. A medical celebrity who fraternized with the Parisian intellectual elite, Charcot studied and treated the large population of insane women housed at the Salpêtrière. Having gained an international reputation for his brilliant work differentiating organic neurological diseases from one another, he eventually became most well-known for his work on hysteria.

Charcot believed that hysteria had an as yet undiscovered biological cause, and that its manifestations followed a predictable course that was universal 'for all countries, for all epochs, for all races' (Shorter, 1993, p.181). Charcot demonstrated—and disseminated—these 'universal' features of hysteria in public lectures held on Tuesdays. These lectures, at first attended only by a small group of fellow professionals, soon began to attract considerable numbers of the public, who attended these dramatic spectacles for entertainment. Charcot selected randomly from a list of patients to be exhibited, thrilling his audience by masterfully finding the signs that would confirm the diagnosis. In fact, Charcot was likely stimulating certain symptoms among his patients through suggestion. 'In Charcot's hands', notes Shorter, 'virtually any patient in these Tuesday lectures could be made into a hysteric' (1993, p.183).

Shorter refers to Charcot's Salpêtrière as a 'crucible of suggestion' (1993, p.182), and the likely role of suggestion in generating classic hysteric symptoms among Salpêtrière patients was noted by contemporaries such as Swiss neurologist Paul Dubois.

> Endowed with the spirit of authority, [Charcot] handled his subjects as he would; and without, perhaps, taking them sufficiently into account, he suggested to them their attitudes and their gestures. Example is contagious even in sickness, and in the great hospitals of Paris, at La Salpetriere, all cases of hysteria resemble each other. At the command of the chief of staff, or of the interns, they being to act like marionettes, or like circus horses accustomed to repeat the same evolutions.... The regularity of the phenomenon is due to the suggestion which the physician, either voluntarily or involuntarily, exercises. (Dubois, 2017, pp.15-16)

Thus, Charcot in part created a new way to be ill. Through his writings and his public demonstrations of hysterical patients, he popularized this new illness and spread its template for others to follow. Charcot-style hysteria remained a common diagnosis throughout much of Europe, but after his death in 1893, the popularity of this presentation began to decline sharply. Viennese psychoanalyst William Stekel noted that 'twenty years after Charcot's death one could not find a single case of hysteria in any of the Paris hospitals' (Shorter, 1993, p.198).

The notion of being transgender could not exist without doctors. Transgender was not a category one could imagine for oneself until medical and surgical advances created transition. Children who were unhappy with their gender were the concern of a tiny minority of specialists until a medical pathway opened up to treat them with the advent of the use of puberty blockers for gender dysphoria. The availability of the treatment appears to have essentially created the demand, and the edges of the category have creeped ever outward. What was once an extraordinary measure for someone suffering greatly from a rare condition has become commonplace.

The use of puberty blockers was introduced to the US by pediatric endocrinologist Norman Spack of Boston Children's Hospital, making him one of the chief popularizers of the idea of the transgender child. Spack first learned about the Dutch puberty blocking protocol while attending a conference in Europe in the early 2000s. He told *The New York Times* of his reaction: 'I was salivating ... I said we had to do this' (Hartocollis, 2015). Spack did indeed go on to 'do this', opening his Gender Management Services Clinic at the Children's Hospital in Boston in 2007. The model quickly spread, as clinicians trained by Spack went on to open clinics of their own in other parts of the US (English, 2011). There are now around 40 such clinics in the United States (Human Rights Campaign). At first used only for extreme cases, the use of blockers has increased dramatically since Spack introduced them in the US (Lopez et al, 2018). Spack was instrumental in propagating the idea of the transgender child in another way. In 2009, he helped draft the Endocrine Society guidelines that recommended commencing puberty blockers in early puberty, followed by a lifetime of cross sex hormones (Ruttiman, 2013).

Spack spread the idea of the transgender child into the symptom pool in a third way. He became that thing which many hysterias have—a doctor celebrity. While Charcot had his Tuesday lectures, Norman Spack had his 21st century equivalent—the Ted Talk. Spack's 2013 talk has been viewed nearly 1.3 million times and has been transcribed into 27 languages, according to the TED website. Here Spack disseminates the template for being a transgender child, which includes the following elements: if a child says they are 'in the absolute wrong body' by early adolescence, they are 'extremely unlikely' to change their minds; untreated transgender people are at a high risk for suicide; kids who go through their natal puberty will 'fall apart;' puberty blockers are a benign intervention that 'buy time;' and kids can definitively 'affirm their gender' even as toddlers (Spack, 2013). The statements that Spack makes in this video reappear again and again in narratives about transgender children.

Just as Charcot could turn anyone into a hysteric, some gender doctors and therapists invested in the narrative of the transgender child let us know that suggestion and medical shaping are at work today just as they were at the Salpêtrière. In the following transcription of a recording of a presentation at a 2017 medical conference, Johanna Olson Kennedy shares an anecdote about an eight-year-old natal female who was brought to her because she was wearing a boy's clothing and hairstyle:

> So at one point, I said to the kid, 'so do you think that you're a girl or a boy? And this kid was like... I could just see, there was, like, this confusion on the kid's face. Like, 'actually I never really thought about that.' And so this kid said, 'well, I'm a girl, 'cause I have this body.'
> And I said, '...Do you ever eat pop tarts?' And the kid was like, oh, of course. And I said, 'well you know how they come in that foil packet?' Yes. 'Well, what if there was a strawberry pop tart in a foil packet, in a box that said 'Cinnamon Pop Tarts'? Is it a strawberry pop tart, or a cinnamon pop tart?'
> The kid's like, 'Duh! A strawberry pop tart.' And I was like, 'so...'
> And the kid turned to the mom and said, 'I think I'm a boy and the girl's covering me up.' (4thwavenow.com, 2017)

Here we witness a physician suggesting to a vulnerable young person a new way to think about herself that the patient herself had not considered before. The child is being unconsciously coached into formulating her experience along the lines suggested to her by the authority of the doctor.

Unhappy, vulnerable patients

Charcot had vulnerable young women who enacted the prescribed script of hysteria on command. Augustine, admitted to the Salpêtrière at age 15 ½, was one such famous patient. She was immortalized in widely distributed photographs of her in the midst of hysteric fits (Shorter, 1993).

It would be difficult to select a single patient responsible for the adoption of trans as a go-to means of communicating distress in our current culture. In 2007, Barbara Walters featured then six-year-old Jazz Jennings, a transgender girl who would go on to become the face of the transgender child in the United States. With frequent media appearances, a children's picture book, and acne treatment sponsorship, Jazz has often been in the spotlight. In 2015, the television network TLC began running a reality show featuring Jazz and her family that documents Jazz's medical transition.

Also in 2015, Bruce Jenner came out publicly as a transgender woman named Caitlyn. Her announcement was followed by intensive media coverage, including a celebrated *Vanity Fair* cover. Janet Mock, Chaz Bono, and Laverne Cox are several other high-profile transgender celebrities. These have helped frame adopting a trans identity as not merely a legitimate option in the symptom pool, but as a valorized one. Being transgender is now a credible way to signal distress, and it also allows the sufferer to transcend the ordinary, to become something special and unique.

British YouTube star and author Alex Bertie was a teenage lesbian who had had no gender discomfort as a girl, although she liked boyish things such as video games, according to a profile of Bertie in *The Times*. At age 13, Bertie became aware that she was attracted to her best friend, a female. The two started dating openly, and homophobic bullying ensued. 'I was known as the 'weird lesbian girl' and nobody would speak to me', Bertie told *The Times* (Turner, 2017).

Over the next three years, the bullying was relentless. She became accustomed to taunts of 'lezzer' and 'you're a boy.' Bertie became troubled and began self-harming. To help herself cope, she set up a YouTube channel and made videos about her life (Turner, 2017).

At last, Alex confided about the bullying and self-harm to a teacher, and the school set up an appointment for Alex to meet with an LGBT group called Over the Rainbow. At the meeting, Alex was given information about transgender issues such as photos of transgender celebrities. Resources were also provided on how to start the NHS process of transition. The meeting and information helped Alex to realize that 'I'm supposed to be male' (Turner, 2017).

Alex began to document his transition on his YouTube channel—first social, and then medical. He began taking testosterone at 19 and had a double mastectomy at 21. He quickly developed a devoted YouTube following which, at the time of writing, exceeds 300,000 people (Bertie, 2018). In 2017, Bertie published a memoir entitled 'Trans Mission: My Quest to a Beard.' *The Times* notes that Bertie is now a 'trans poster boy', and that his book is a 'how-to manual for confused girls with an appendix on how to make your own breast binder' (Turner, 2017). At a vulnerable moment, Bertie was offered the template of transgender, and perhaps saw it as a way to legitimize and exalt his loneliness and suffering while also offering a clear path to transformation and treatment, and now his book and YouTube channel have become part of the feedback loop, making the template available to many more.

Supportive cultural environments

Bertie's experience as reported in *The Times* suggests a troubling possibility—that the template for trans was offered by well-meaning school officials to a young lesbian struggling with isolation and rejection. I cannot help wondering whether Bertie might have arrived at a different understanding of himself if Over the Rainbow had given him different resources. The information provided to Bertie that day encouraged him to interpret his distress in a certain manner—one that would eventually lead him to feel he needed hormones and surgery. This is one of the many examples of how certain cultural environments such as schools and outreach programs can cultivate and disseminate a hysterical template such as gender dysphoria.

The template of Charcot's hysteria was propagated through his students, who took his teachings with them to other centers. It was spread by colorful accounts in the newspapers and magazines of the day. And it was disseminated through portrayals of the symptoms that appeared in literature. Charcot-style hysteria became so widely known that it became commonplace for ordinary people to enact its symptoms. For example, the French writer Edmond de Goncourt wrote in his diary in 1889 that 'it is truly a bit unsettling … how society women are carrying on right now. They all seem like the *hysteriques* of the Salpetriere, let loose by Charcot upon the world' (cited in Shorter, 1993, 188).

While Charcot had his Tuesday lectures, today's propagators of the latest template have a much more powerful dissemination tool—the internet. Sites such as YouTube and Tumblr may be responsible for more trans teens than nearly any other medium. A descriptive study of parent reports regarding teens who suddenly declared themselves transgender found that around 65% had had an increase in social media use around the time they came out as trans (Littman, 2018). This research observation is backed up by anecdotal evidence presented in countless news reports about transgender children, who frequently note that they first entertained the idea that they might be trans after learning about the concept online. The following quote is from a newspaper article about an eight-year-old who identifies as a boy.

> 'It's hard to explain,' Joe says in his living room as he plays with his family's 3-month-old rescue dog, Oreo. 'You know I actually have a disease to make me feel like I'm a boy...'
> 'No, that's not a disease,' Kristie [his mom] tells him.
> 'No it is!' Joe replies in what is becoming a heated debate inside the apartment.
> 'Where are you getting your information from, huh?' Kristie asks.
> **'I'm dead serious!' Joe says. 'I looked on YouTube!'** (McDonald, 2017, my emphasis)

Joe learned that it was possible to be transgender from YouTube, and his story was published in a newspaper, further distributing the template for others to find—another example of the feedback loop. Hardly a week goes by without a major media outlet carrying a story about a transgender child or teen. Such media coverage is another important way that being transgender gains traction. Watters made a similar observation about the spread of anorexia when he noted that each news article or television show that features the condition increases 'incrementally ... the gravitational pull of the disorder on the unconscious minds of the population' (Watters, 2011, p.49).

Indeed, Hannah Bruch, one of the leading authorities on eating disorders, noted the stark contrast between patients who developed anorexia before and after the recent popularization of the disease in the 1970s. The latter patients often evidenced peer and media influence. The two populations were quite distinct in their clinical presentation. Bruch called those who developed the disease after it became a widely known cultural phenomenon 'me too' anorexics, and noted that most knew about the illness, 'knew someone who had it... [or] deliberately 'tried it out' after having watched a TV program' (Bruch, 1985, p.11).

Media reports about eating disorders tended to portray anorexics as suffering in a way that was dramatic and worthy of compassion and attention. News reports about transgender children tend to take things one step further, painting trans kids as benighted, but also as authentic, brave, and at the cultural vanguard. The 2017 issue of *National Geographic* that focused on gender was titled 'The Gender Revolution' (Goldberg). Who would want to be left behind?

Social factors that valorize a condition, confer special status on the victim and her family, or link the condition with progressive social movements can create further fertile ground for a contagion. Watters notes that, because anorexia came to be connected with a feminist critique of culture, this 'unintentionally glamorized the disease and elevated the social role of the sufferer' (2011, p.58). This is dangerous because the exalted status bestowed upon those with such a condition encourages more people to adopt it.

Numerous glowing media reports about brave trans kids and their heroically supportive families offer celebrity status to both, explaining perhaps why so many parents of trans kids have public blogs that document their parenting experience with a trans child. An article that appeared in the online outlet *Vice* is typical of the hyperbole reserved for supportive parents. Entitled 'How the Mothers of Transgender Children are Changing the World', the piece notes that mothers of trans-identified kids often go on to become activists and advocates (Tourjee, 2016).

A blogging mom of a trans child wrote a post entitled 'Do You Want to Be Me?' She concludes that, yes, other parents might want to be her, because the life of a parent of a trans kid is in some respects genuinely exalted. 'I guess we look inspirational. I've realised it's because we are.... When you skim across the surface our lives look glamorous with a soupcon of drama.... We are thousands of stories of ordinary people on an extraordinary parenting journey. What binds us is that we are testament to the power of love' (Parenting Jeremy – A Gender Journey). Parents of trans kids are seen as heroic and worthy of emulation.

The same valorization that parents receive is also conferred on trans kids themselves. In February 2018, *The New York Times* reported on the GenderCool project, which was formed to celebrate gender diverse youth. It calls its young members 'champions', and showcases them on national television programs such as 'Megyn Kelly Today' (Caron, 2018). When a condition is seen as glamorous, our cultural fascination can become part of the feedback loop, further fueling the contagion.

What can be done?

Toward the end of his life, Charcot began to accept that hysterical symptoms might in fact be psychogenic in origin. Charcot's student Sigmund Freud would take this idea further by linking hysteria to repressed sexual fantasies and desires. Freud's own student Jung would take a psychological view of symptoms even further, seeing them as the psyche's expression of its own attempt to heal itself. As Charcot moved toward a psychological understanding of hysteria, his patients tended to drop these symptoms in favor of those with more credibility (Shorter, 1993).

How can we interrupt the feedback loop that is encouraging thousands of young people to seek body altering treatments? The medical and mental health fields must guard against becoming entranced by exotic new conditions, such that, in our excitement, we unconsciously reinforce maladaptive behaviors. We ought to continue to do the hard, psychological work of treating the underlying issues that affect our patients, while taking care not to subtly encourage the expression of hysterical symptoms. And we ought to insist on patient responsibility.

We must also give the unconscious its due. Showalter encourages us to allow 'more space for the mysteries of human emotions', rather than seeing experiences through a narrow, medicalized lens (1997, 11). I am certain that Jung would agree. The messages that our unconscious seeks to convey are ultimately mysterious, but if we listen to them with curiosity and reverence, our understanding of ourselves and others will continue to deepen. If hysteria is the psyche's way of speaking to us, we would do well to listen.

References

4thwavenow.com (2017, July 25). 'I just gave him the language': Top gender doc uses pop tart analogy to persuade 8-year-old girl she's really a boy. *4thWaveNow*. Retrieved from https://4thwavenow.com/2017/07/23/i-just-gave-him-the-language-top-gender-doc-uses-pop-tart-analogy-to-persuade-8-year-old-girl-shes-really-a-boy/

Arcelus, J., Mitchell, A. J., Wales, J., & Nielsen, S. (2011). Mortality Rates in Patients with Anorexia Nervosa and Other Eating Disorders. *Archives of General Psychiatry, 68*(7), 724-731. doi:10.1001/archgenpsychiatry.2011.74

Bertie, A. (2018). The Real Alex Bertie [video file]. Retrieved from https://www.youtube.com/user/TheRealJazzBertie

Bruch, H. (1985). Four Decades of Eating Disorders. In D. M. Garner & P. E. Garfinkel (Eds.), *Handbook of psychotherapy for anorexia nervosa and bulimia* (pp.7-18). New York: Guilford Press.

Caron, C. (2018, February 20). These Transgender Children Say They're Thriving. They Want to Help Others do the Same. *The New York Times*. Retrieved from https://www.nytimes.com/

Dubois, P. (2014). *The Psychic Treatment of Nervous Disorder: The psychoneuroses and their moral treatment ... (classic reprint)*. Amazon Digital Services.

English, B. (2011, December 11). Led by the Child Who Simply Knew. *The Boston Globe*. Retrieved from https://www2.bostonglobe.com/metro/2011/12/11/led-child-who-simply-knew/SsH1U9Pn9JKArTiumZdxaL/story.html

Feder, S., Isserlin, L., Seale, E., Hammond, N., & Norris, M. L. (2017). Exploring the association between eating disorders and gender dysphoria in youth. *Eating Disorders, 25*(4), 310-317. doi:10.1080/10640266.2017.1297112

Goldberg, S. (2017, January). Gender Revolution. *National Geographic*.

Gordon, R. A. (2000). *Eating disorders: Anatomy of a social epidemic*. Oxford: Blackwell.

Hartocollis, A. (2015, June 17). The New Girl in School: Transgender Surgery at 18. *The New York Times*, p.A1. Retrieved from https://www.nytimes.com/

Human Rights Campaign. (n.d.). Clinical Care for Gender-Expansive Children & Adolescents. Retrieved from https://www.hrc.org/resources/interactive-map-clinical-care-programs-for-gender-nonconforming-childr.

Inglis-Arkell, E. (2014) The "Glass Delusion" Was the Most Popular Madness of the Middle Ages Retrieved from https://io9.gizmodo.com/the-glass-delusion-was-the-most-popular-madness-of-th-1636228483.

Jung, C. G. (1961). *Collected Works of C.G. Jung, Volume 4, Freud and psychoanalysis* (R. Hull, Trans.). Princeton, NJ: Princeton University Press.

—. (1967). *Collected Works of C.G. Jung, Volume 13, Alchemical Studies* (G. Adler, Ed.; R. F. Hull, Trans.). Princeton: Princeton University Press.

Jung, C. G., & Adler, G. (1970). *Collected Works of C.G. Jung, Volume 10, Civilization in Transition* (R. F. Hull, Trans.). Princeton: Princeton University Press.

Littman, L. (2018). Rapid-onset gender dysphoria in adolescents and young adults: A study of parental reports. *PlosOne, 13*(8). doi:10.1371/journal.pone.0202330

Lopez, C. M., Solomon, D., Boulware, S. D., & Christison-Lagay, E. R. (2018). Trends in the use of puberty blockers among transgender children in the United States. *Journal of Pediatric Endocrinology and Metabolism, 31*(6), 665-670. doi:10.1515/jpem-2018-0048

Marchiano, L. (2018). The Language of the Psyche: Symptoms as Symbols. In H. Brunskell-Evans and M. Moore (Eds.), *Transgender Children and Young People: Born in your own body* (pp.41-63). Newcastle upon Tyne: Cambridge Scholars Publishing.

McDonald, C. W. (2017, January 7). After ejection from Cub Scouts, transgender boy tells his story. *NJ.com*. Retrieved from https://www.nj.com/

Parenting Jeremy - a Gender Journey (2016, August 28). Do You Want to Be Me? Retrieved from https://parentingjeremy.com/2016/08/28/do-you-want-to-be-me/.

Ruttiman, J., PhD. (2013, January). Blocking Puberty in Transgender Youth. *Endocrine News*.

Scull, A. T. (2011). *Hysteria: The disturbing history*. Oxford: Oxford University Press.

Selmakuvar, S. (2014). Dr. James Hollis: Why Is Myth Important in Our Modern Lives? Retrieved from http://www.jungfl.org/dr-james-hollis-why-is-myth-important-in-our-modern-lives/

Shorter, E. (1993). *From Paralysis to Fatigue: A History of Psychosomatic Illness in the Modern Era*. New York, NY: Free Press.

Showalter, E. (1997). *Hysteries: Hysterical epidemics and mass media*. New York, NY: Columbia University Press.

Spack, N. (2013). *How I help transgender teens become who they want to be*. Retrieved from https://www.ted.com/talks/norman_spack_how_i_help_transgender_teens_become_who_they_want_to_be

Tourjee, D. (2016, August 23). How the Mothers of Transgender Children are Changing the World. *Vice*. Retrieved from https://broadly.vice.com/

Turner, J. (2017, November 11). Meet Alex Bertie, the transgender poster boy. *The Times*. Retrieved from https://www.thetimes.co.uk/

Watters, E. (2011). *Crazy Like Us: The Globalization of the American Psyche*. New York, NY: Free Press.

Weir, E. (2005). Mass sociogenic illness. *Canadian Medical Association Journal, 172*(1), 36.

CHAPTER FOUR

PSYCHIATRY AND THE ETHICAL LIMITS OF GENDER-AFFIRMING CARE

ROBERTO D'ANGELO

> There is much more to children than what they say. We owe to them a deeper listening than a literal one ... [or] we will miss whatever new story they are telling or protest they are making. (Schwartz, 2012)

Brunskell-Evans (2018) has argued that the concept of the transgender child is a newly created concept, an entirely new way of thinking about what it means to be human, that previously did not exist. From its emergence as a new idea within late 20th century psychiatry, it is now an established focus of mainstream psychiatric care, as evidenced by the large number of specialised childhood gender clinics in the USA (Hsieh and Leininger, 2014). The concept of the transgender child and gender-affirming therapies, including medical and surgical transition, are an extension of the mechanistic, technological paradigm of modern biomedical psychiatry. Biomedical psychiatry's focus on locating defects in the individual mind/body serves to obliterate our awareness of the complex environmental factors that shape our subjectivity and that generate human suffering, including gender dysphoria in young people. This raises questions about the ethical limitations of psychiatry's gender-affirming treatment paradigm, as its narrow focus on correcting a mismatched body and brain fails to attend to the interpersonal, family and social context that may be generating the child's gender distress. The psychiatric medicalisation of childhood gender distress concretises 'the story' gender-troubled children are telling and by focussing on bodily alteration, it collapses space for imagining new and truly liberating possibilities.

Gender-affirming care as an extension of biomedical psychiatry

21st century psychiatry operates increasingly within a medicalised and neurobiological paradigm, firmly grounded in the physical sciences. This biomedical psychiatry is situated in a discourse in which human differences, suffering and illness are understood to derive from neurological or neurochemical substrates in the brains of individuals. Advancements in neuroscience, which are a central part of this project, attempt to locate personality, gender identity and other individual differences in brain structure. Similarly, emotional and psychological problems are understood as manifestations of dysfunctions in neurotransmitter systems or alterations in specific brain regions. Psychiatry's conviction is that suffering is due to a defect in the individual mind and/or body and that relief is to be found in improved diagnostic technologies and corrective medical and pharmacological treatments.

This paradigm is sometimes referred to as the 'technological paradigm' (Bracken et al., 2014). Notwithstanding the contributions it has made to our understanding of neurobiology, it presents us with a profoundly de-contextualised and mechanistic way of understanding ourselves. Further, it offers only a very circumscribed framework for how we understand individual suffering and distress. It is based on a positivist epistemology that understands mental illness to be the result of faulty processes within the individual that are *not* context-related. According to this understanding, whether we thrive, or whether we suffer, is largely a product of our genes, our biology, our brain, and our neurotransmitter systems (Capra, 1982). The decontextualisation that this entails de-emphasises or even erases the contribution of our lived relational history and current interpersonal relations to our ongoing experience at any point in time. It also ignores the power of social, political and economic forces to shape individual experience, generate suffering or even cause illness.

This paradigm has profoundly influenced how we think about and respond to gender variant children. In fact, it might be argued that the notion of the transgender child has emerged in parallel with the rise of biomedical psychiatry and perhaps that it could only emerge in such a paradigm. The way the phenomenon is understood and the primary, endorsed treatment approach, known as gender-affirming care, are extensions of psychiatry's biological and atomised way of understanding human beings. Gender distress is seen as something occurring exclusively *inside* the individual. Children experiencing gender dysphoria are understood as having a mismatch between their gender identity, presumed to reside in neural structures or networks, and the sex of their body. The solution to gender dysphoria involves correcting the mind/body mismatch. The problem is a seen as a physical, medical one, located within the individual mind/body, and therefore requires a concrete solution, frequently involving biological treatments.

The biomedical framework considers the interpersonal and social context in which the child develops and experiences gender-related distress to be of little relevance, beyond the degree to which the family and culture are supportive (or not) of the child's gendered experience. The child's gender identity is conceptualised as existing *a priori,* as though it is an essential quality of the child that is beyond the reach of social and environmental influences. How gender dysphoria may emerge in response to family dynamics, traumatic experiences such as emotional or sexual abuse, or social issues including how masculinity and femininity are constructed and regulated in our culture is not relevant to this formulation. Rather, gender identity is seen as a fixed, internal quality of the individual, or a fundamental core essence that reflects an essential truth about the child:

> The true gender self begins as the kernel of gender identity that is there from birth, residing within us in a complex of chromosomes, gonads, hormones, hormone receptors, genitalia, secondary sex characteristics, but most importantly in our brain and mind….its center always remains our own personal possession, driven from within rather than from without. (Ehrensaft, 2012, p.341)

This has been reinforced by authorities in the newly emerging field of childhood gender medicine, who assert that children can know that they are transgender from a very young age (Ehrensaft, 2014; Olson et al., 2016), and that transgender children represent variants of normal human developmental trajectories (de Vries and Cohen-Kettenis, 2012). Gender and gender identity have become reified by biological psychiatry, as 'things' located in the brain, with gender identity even described as being 'indelible from birth' (Department of Health, 2008). Transgender children are understood as having a 'female brain in a male body', or vice versa, despite the fact that, to date, there is no convincing evidence that that this is the case (Guillamon, Junque & Gomez-Gil, 2016). Because gender identity is seen as a kind of immutable psychic bedrock, psychiatric orthodoxy posits that the ethical and progressive way to respond is to facilitate the expression of the child's subjective, 'true' gender identity. This notion has rapidly established itself in the cultural psyche beyond health care, and now permeates multiple aspects of our culture: the media, education, and legislation. The dominant discourse involves promoting the acceptance of transgender children and ensuring earlier and easier access to the recommended treatments for gender dysphoria, namely social and medical transition.

Our response to gender variant children has become almost exclusively determined by this approach, despite the reality that the outcome data for social and medical transition is often of poor quality and certainly not conclusive (Murad et al., 2010; D'Angelo, 2018). Psychological therapies are considered helpful only as adjuncts to transition or to treat comorbid problems (World Professional Association for Transgender Health, 2009). Psychological treatment for gender dysphoria itself is assumed to be unhelpful (Lawrence, 2014; World Professional Association for Transgender Health, 2009), however no data actually backs up this claim. Psychiatry has now abandoned any investigation of the potential effectiveness of psychological therapies as standalone treatments for gender dysphoria. Alternative frameworks for understanding gender dysphoria, such as the one described in this chapter, are dismissed as misguided or framed as transphobic. A recent paper by Littman (2018) suggested that some cases of gender dysphoria were generated by social factors in psychologically troubled teens. The paper was discredited by trans activists and intense lobbying called for the paper to be retracted.

The dominant, accepted psychiatric discourse entails a closing down of space to think about the individual, relational and social meaning of childhood gender dysphoria. On the level of the individual, we do not investigate why this particular child feels they are the gender they feel they are, beyond asking about their preferences for gendered toys, clothes, games, etc. We do not ask what it means in the context of their particular developmental history or their current family and social context to be a man or a woman. Such questions are considered pathologising because they appear to seek out the causes and aetiology of the child's gendered experience, rather than accept it as an innate, 'true' essence. It is concerning that this kind of exploration could potentially contravene many current professional treatment guidelines and place the clinician at risk of accusations of unethical conduct. On the level of culture, we assume that the continuing exponential increase in presentations to childhood gender clinics (Zucker, 2017) is because of a greater social acceptance of this presumably naturally occurring phenomenon. Attempts to understand why the transgender child has emerged as such a pressing and apparently growing phenomenon at this current point in our western history, or to consider the complex social forces that may be in play, are increasingly excluded from psychiatric and social discourse.

Catherine

Catherine received treatment for gender dysphoria informed by the current, medicalised approach and transitioned from female to male. At the age of 18, she underwent a bilateral mastectomy, hysterectomy and bilateral oophorectomy after having been on testosterone for two years. She had begun to experience gender dysphoria around puberty when she increasingly experienced her hips and breasts as intolerable. She had, however, been a deeply unhappy and painfully shy child, with significant interpersonal and emotional issues that predated the onset of her gender dysphoria by many years. Despite this background, psychiatric consultation assessed her as suitable for medical and surgical transition. She had had a total of three contacts with psychiatric services prior to transitioning surgically.

Catherine had been a troubled teen. She had no friends and felt on a deep level that she was unlovable and that no one would ever want to be close to her. Through extensive online research, she formed the conclusion that she was transgender and became convinced that transitioning was the answer to her many complicated emotional and social difficulties. She believed that after transitioning, she would no longer be shy, that people would like her and that her painful sadness would end. None of this occurred and rather than being liberated, she, now he (Josh), became severely depressed and suicidal. Josh's life came to a standstill. He was unable to work or study and eventually became unable to leave the house. Some years later, now completely unable to get on with his life, he reluctantly sought psychotherapy as a last resort.

It became clear that his family and developmental history was given only superficial attention during his contacts with mental health. Through the process of his post-transition, long-term psychotherapy, Josh realised that the pain he was seeking relief from was actually not about gender, but about other issues related to his lived history. Josh began to understand that he had formed a belief that there was something wrong with him in response to a very problematic home environment. Catherine had blamed herself, as many children do in response to dysfunctional or abusive environments (Davies and Frawley, 1994). Her parents were relentlessly critical, emotionally distant and prone to frightening attacks of rage. As a young, self-conscious teen, Catherine's mother repeatedly made disparaging comments about her weight, the size of her thighs and her breasts. Catherine's shame grew as she internalised her mother's contempt for her developing female body. At the same time, her older brother engaged in sexually inappropriate behaviours with her and had no respect for her personal, bodily boundaries. He frequently spoke in a derogatory, sexualised way about women's bodies, especially their breasts. At a young age, Catherine stumbled upon porn sites he had viewed, and found the images frightening and disturbing. Josh came to see that all of these experiences meant that Catherine's experience of her body and gender had been infused with the derogatory and critical attitudes that had permeated her home environment. He also realised how vulnerable, ashamed and alone Catherine was, and how much she had hated herself.

Josh realised that transitioning was a way to eliminate Catherine and begin again. It was a form of internal suicide, intended to eliminate the pain and self-hatred she felt, which had become inextricably linked with femaleness. Psychotherapy involved a very painful process in which he realised that he would probably have been happy with her female body, if he had had access to psychotherapy at a young age. His identification as male, which he held with the strongest conviction for many years, shifted profoundly over time. He eventually no longer identified as male and even considered stopping testosterone. He came to feel that gender had consumed his life and he struggled with grief about the enormous amount of time he felt he had wasted pursuing gender change. The case of Josh exemplifies how even if a young person has apparent certainty about their gender identity, providing gender-affirming care may in fact entail an ethical failure. It also illustrates how gender dysphoria emerges in a context and how gender identity can evolve and change when these issues are comprehensively addressed.

A dynamic systems approach to understanding gender dysphoria

Biomedical psychiatry is arguably an extension of 19th century mechanistic science, which sought to identify essences, or the parts that make up the whole. There have been radical shifts in thinking which challenge this perspective, particularly involving the recognition of complexity, networks and patterns of organisation (Capra and Luisi, 2014). A new framework for understanding natural phenomena, known as non-linear dynamic systems theory, has emerged in recent decades and presents a radically different way to understand living organisms. In this framework, the focus is on the processes by which systems (including living organisms) are ordered, rather than on essences or parts (Thelen, 2005). Systems are irreducibly context-dependent and continually influence and re-organise each other at multiple levels in complex and non-linear ways. This approach is consonant with contemporary psychoanalytic models, which understand people not as mechanistic, isolated 'bio-machines' but as constantly in process (Seligman, 2005). People are conceptualised as dynamic and responsive beings in constant relation with others, social-political systems and their environment. This relation is bidirectional from the first moments of life (Beebe and Lachmann, 1998), meaning that while individuals are constantly influenced by their relations and environment, they also influence and shape them in return.

This approach leads to a way of thinking about gender that is very different to the mainstream psychiatric one. Gender identity would be seen as an emergent phenomenon, arising through the complex interplay of multiple systems, including any biological ones. Gender identity is not a 'thing' but a 'pattern in time' (Fausto-Sterling, 2012) that is continually being influenced by the relational and cultural surround and also shaping the responses of those individuals and systems it comes into contact with. Dynamic systems theory posits that novel inputs will have impacts that cannot always be predicted, and that small changes can sometimes result in dramatic reorganisations (Thelen, 2005). Gender identity is thus understood as a pattern that emerges at the interface of a potentially limitless number of influences. These might include temperament; the sex of the body; neurological variations; parent-infant interactions (Fausto-Sterling et al., 2015), including the parents' conscious and unconscious responses to the infant's sexed body; the psychology of both parents, including both strengths and pathology; trauma experienced by the child or a parent or the family, including death, physical or sexual abuse; family dynamics, both healthy and dysfunctional; peer relations, advertising, social media, popular culture, and pornography; the influence of religious or secular ideologies; the structure of cultural discourses about men and women influenced by patriarchy, feminism, queer theory and so on; economic and social conditions, including poverty, migration, and marginalisation; or political ideologies such as capitalism, democracy, and dictatorships. This list of potential interacting systems is virtually endless.

Because of the fundamental assumption that individuals and systems constantly affect each other in a bidirectional and non-linear way, dynamic systems theory assumes that treatment interventions will become a part of the child's context. This means that they will have the power to influence and reshape the trajectory of the child's future development, including how they will come to understand themselves and the choices they will make. An approach that affirms the child's stated gender is not neutral: it is a very powerful input into the system that will have an effect. This appears to be borne out by data which indicates that early social transition reduces the likelihood that a child will desist (Steensma et al., 2013). Equally, an exploratory approach which seeks to develop a detailed understanding of the child's gendered and lived experience within multiple systems will have its own impacts. Potentially, by helping the child develop an increased understanding, or insight (Jennissen et al., 2018), into why they have come to feel the way they do, the child and the family will be in a better position to decide whether social and medical transition is really the best way to deal with the situation. Therapeutic intervention with the individual and/or the family may even result in a change in the individual's gendered experience, as in the case of Catherine. The outcome may then be that medical and surgical interventions are no longer needed or desired.

Identifying gender dysphoria as a 'thing' makes no sense in this conceptualisation, as gender dysphoria cannot be understood without understanding the context in which it arises. Rather than seeing the child as *in actuality* being born in the wrong body, this kind of approach would seek to understand in what context(s) this particular child has come to *feel* that they have the wrong body. This would entail detailed exploration of the many factors already outlined. The therapeutic process it entails is one which attempts to be open to all possibilities and the possibility of change, without making any *prima facie* assumptions about the immutability of gender or the veracity of children's gender claims. At the same time, it does not set about to change the child's gender identity but seeks to understand it as a 'pattern in time'. This kind of exploration is time-intensive and is itself an evolving process. It is not possible to work at this level of complexity and breadth in one or two sessions, the sum total of Catherine's contact with mental health.

This raises serious questions about the adequacy of the current psychiatric paradigm and about what constitutes an ethical response to children and young people expressing gender discordance or distress. Whilst it appears that affirming the child's stated gender ameliorates their suffering (de Vries et al., 2014; Olson et al., 2016), is it possible that it might leave the foundational and potentially serious difficulties and issues in the child's life unnoticed and untreated? Catherine had been assessed and treated within a framework that fails to inquire about the context and meaning of individual experience, particularly in relation to gender. She received standard, gender-affirming care in line with WPATH standards. Her gender identification as male was responded to as though it was a kind of bedrock that need not be questioned or explored. As she underwent medical and surgical transition, the very real issues in her life, including her problematic history, which included emotional and sexual abuse, and its consequences for her capacity to connect and sustain closeness, fell further and further out of focus. Ultimately, it was as though these issues had been forgotten about by both Catherine and her treating physicians. When Catherine first consulted me, some years after her transition, severely depressed, completely isolated and unable to work, she had no idea why she felt so bad. All she knew was that she had been counting on transitioning to transform her life and it had not.

Mika

Mika is a 12-year-old girl who had been referred to a therapist at the age of 10 for the treatment of anxiety. She had begun to experience separation anxiety when leaving the home, in particular when leaving her parents, fearing that something terrible would happen to them. She was a shy child who did well at school but struggled to make friends. There had been no previous behaviour suggestive of gender-nonconformity and she had not expressed any concerns about her gender prior to this. Her therapist had been helping her with strategies to manage her anxiety, with little effect. At the age of 12, she told her therapist that she felt that she was a boy. Her therapist was accepting and supportive and asked to meet with her parents. The therapist told Mika's parents that their child was transgender and that they should allow her to socially transition as soon as possible. The therapist believed that Mika's anxiety was secondary to Mika being transgender and that it would resolve when Mika was able to express her true gender and be affirmed in it by her family and friends.

Unfortunately, Mika continued to be anxious and in fact became more distressed after socially transitioning, despite stating that she now felt more comfortable as a boy. Her parents became concerned and wondered if they should seek a second opinion. At this consultation, Mika explained how she came to realise she was a boy. From around the time of puberty, she felt increasingly alienated and unable to relate to the other girls. She did not like wearing dresses or skirts and was not interested in make-up or dating boys. She began watching YouTube videos of trans guys and other trans teens who spoke about how transitioning had relieved them of their anxiety and sadness. Mika concluded that the reason she felt anxious and sad was the same as for the people that she was watching on her screen: she was also trans.

After a brief improvement, Mika's fears that something terrible would happen to her family intensified. Exploration of the family system revealed that the marital relationship had been problematic for some years. The onset of Mika's anxiety coincided temporally with the onset of marital difficulties. This culminated in a serious marital crisis resulting in a 6-month separation. Mika's announcement that she was a boy coincided with the end of the separation and reunification of the family. Both parents believed they had protected their children from knowing about the marital difficulties. However, it emerged that Mika's father had clearly been very depressed for some time prior to the separation, spending long periods withdrawn from the family or in bed. Both parents were in fact depressed and reported having felt intermittently suicidal. They had tried to maintain the appearance of a happy and functional family without addressing the difficulties in their relationship.

Even though none of this was disclosed to Mika, the marital impasse profoundly shaped Mika's emotional and relational context. Mika had developed a sense of foreboding, a feeling that something was wrong, leading to fears that her parents might die. She did not know why and sometimes wondered if she was crazy. It turned out that she was in fact responding to very real issues in the family system but had no way of understanding her feelings, as the truth was being concealed from her. There was a profound mystification (Laing and Esterson, 1964; Laing, 1985), or gaslighting, going on within the family. As children often do, Mika blamed herself and assumed there was something wrong with her. This became interwoven with Mika's sense that she was different to the other girls (probably due to an emerging same-sex attraction). Trans became a concrete, tangible way for her to make sense of how unhappy she felt and also why she was different. The many triumphant YouTube videos she watched, in which trans guys shared their heroic stories, gave her hope and a way out of her despair. The family was referred for family therapy and both parents began individual psychotherapy.

Ethical limitations of gender-affirming care

i. The erasure of personal meaning and relational context

Mika's case illustrates how working within a systemic or contextual clinical frame leads to conceptualising gender distress as an emergent phenomenon within a complex system. This case highlights the impact that differing theoretical understandings of gender dysphoria will have on how the child's problems are understood and also on what is considered to be the focus of treatment (Schwartz, 2012; Thelen, 2005). Mika's therapist operated within a gender-affirming frame and her approach makes sense within the essentialist, gender-as-concrete-reality paradigm. However, her formulation and the resulting advice she gave to the family bypassed the very real issues in Mika's life, particularly the serious family dysfunction. The therapist's framing of Mika's anxiety as secondary to gender dysphoria also reinforced the mystification that Mika was experiencing. Rather than helping Mika understand the real, contextual reasons for her feelings of anxiety and foreboding, her therapist cemented the idea that Mika felt so bad because there was something wrong with her (she had the wrong body).

Whilst her therapist believed she was responding ethically, there are important questions here about whether an approach that does not address serious systemic (family) and psychological (individual and family) issues in fact fails the child and their family. Does such an approach really help children claim their authentic selves or does it produce a mystification with the result that they may never come to really know themselves? Would Mika's problems remain forever unarticulated and unformulated, as the psychiatric system narrowed its attention solely on her gender and her body? Or would Mika eventually come to realise why she was so unhappy, that it actually had very little to do with gender, just like Catherine did when transitioning failed to be the solution she had hoped for? This case illustrates how the endorsed treatment paradigm can give gender dysphoria a kind of special status in which it 'trumps' all other issues, and in which it is de-linked from contextual factors. Gender dysphoria becomes the primary and urgent focus of clinical attention, with other emotional issues being viewed as coincidental comorbid conditions, or as secondary responses to the distress caused by being born in the wrong body.

Gender-affirming care is promoted by treatment guidelines as the singular ethical approach to the treatment of gender dysphoria (Telfer et al., 2018). Viewed through a contextual lens, such an assertion is simplistic as it fails to take into account the meaning of the gender distress in each individual case, and the relational context in which it arises. An affirming approach may indeed be helpful in some cases, however it may entail a serious ethical failure in others. Indeed, providing gender-affirming care *without* exploring the context in which the gender distress is emerging is arguably unethical. The cases of Catherine and Mika are such cases, in which the gender-affirming paradigm obscured the primary issues causing their distress and generated mystification rather than helping them connect to what was authentic. A dynamic systems, context-sensitive approach involves listening very carefully to the child's concerns, trying to understand what the child may be struggling with and attempting to express, rather than taking their gender expressions at face value. Critics argue that anything other than affirmation is pathologising. However, dynamic systems theory eschews single-cause explanations and instead argues that 'everything counts', i.e. multiple interacting systems come together in non-linear ways to produce a particular outcome (Thelen, 2005). A dynamic systems approach involves a much more complex sensibility that resists *either* pathologising *or* affirming, both of which fail to take into account the complexity of individual children's lives.

Ehrensaft (2012, 2014) asserts that children know their true gender self from a young age and that we should listen to them as they know what they need. There is a certain naiveté to this argument in that it assumes that what the child wants is best for the child. A dynamic systems-informed psychoanalytic theory makes no such assumption. Its starting point is that we cannot presume to know the meaning or underlying motivation of any particular desire or behaviour. It seeks to understand the unique meaning and function of gendered experience in each individual case. The desire to change gender could be a generative expression of the child's true self, or it could be an emergent feature of a distressed family system. Such an approach would also be open to the possibility that the desire to transition may be destructive, a way to kill off part of the self, as in the case of Catherine. Its specific meaning can only be established through careful and detailed exploration. A recent article about trans youth (Singal, 2018) that appeared on the cover of the *Atlantic* illustrates this complexity through numerous case examples in which transitioning was sometimes helpful and sometimes not. The individuals in the article for whom transitioning actually did not address their difficulties had been just as convinced that it was what they needed as those with good outcomes.

ii. The erasure of the social context

Just as biomedical psychiatry's understanding of gender distress edits out the many potential sources of turmoil and difficulty in young people's family and relational systems, it also entails an erasure of how our social environment may be implicated in the emergence of gender dysphoria. Despite the considerable progress we have made as a society towards addressing patriarchy, gender continues to be at the heart of serious structural problems. Women continue to experience considerable disadvantages in employment, education, and legislative power, and the gender pay gap persists (World Economic Forum, 2017). Domestic violence is widespread and women are far more likely to be victims of violent crimes committed by their intimate partners than men (Violence Policy Center, 2017). The 'me too' movement forces us to acknowledge the disturbing pervasiveness of women's vulnerability to sexual exploitation and assault. The victims of sex trafficking are primarily women and girls. These issues are part of the gendered social landscape into which children are 'thrown'. Gender-affirming care fails to consider the impact of the gendered social context on gender-dysphoric children beyond acknowledging that trans people face considerable stigma. Further, whilst gender-affirming care is presented as a radical challenge to gender prescriptions, its narrow focus on gender identity and the body arguably does very little to address these disturbing social realities.

The focus on gender dysphoria as a reified problem within the individual also diverts our attention from many other pressing social issues that profoundly affect our lives. Western cultures still fail to effectively address the most devastating issues that affect children on a huge scale, such as sexual abuse occurring within families and institutional settings, economic and educational disadvantage, and poor access to health care. Adults and children alike are troubled by pervasive issues such as the growing divide between the privileged and the poor, the rise of ultra-right-wing politics, increasing political instability and the looming threat of environmental degradation and climate change. Queer theorists argue that neoliberal capitalism generates feelings of inadequacy in individuals to lock them into a lifetime of endless production and consumption (Ruti, 2017).

When we encourage children to narrow their focus on discovering their true gender self, do we at the same time encourage them *not* to think about more profound reasons for why they may feel so alienated, or despairing or disempowered? Are we helping children understand themselves and their world by helping them express their 'true' gender, or are we mystifying them by promising liberation through gender change whilst diverting their attention from crucial social realities? Further, are we participating in a cultural collusion to ignore the most pressing problems that face us as a culture, and distracting ourselves with the project of reconfiguring our appearances, bodies and gendered behaviours? And does this represent a radical challenge to gender norms, or is it simply a new iteration of consumer capitalism, which promotes an endless array of possibilities for self-reinvention (Ruti, 2017)?

Specifically, in relation to gender, children with gender dysphoria feel that they do not fit how the culture expects them to be as a boy or a girl. Are gender-dysphoric children responding to the way rigid gender constructions still saturate our culture at multiple levels, via advertising, education, popular culture, social media, porn and so on? Do they feel increasingly suffocated by the limited and limiting options available to them within a gendered world (Plan Int., 2018)? Some have observed that even within subcultures that have traditionally championed gender non-conformity, rigid gender constructions increasingly define what is acceptable and desirable (Jeffreys, 2014). There is increasing social pressure on gay men, for example, to conform to a hypermasculine gender performance, and the requirement 'no-fems' has become a prevalent feature of online hookup and dating profiles. Free pornography platforms mean that images of women being sexually subjugated by men increasingly shape how young people understand sexuality and sex relations (Jones, 2018).

Is the distress of gender dysphoria an attempt by young people to communicate that living within current gender regulation is untenable? And is the inherent rejection of 'assigned gender' an attempt to imagine a different way to live? What this suggests is that the acceptance of children's gender claims as literal statements could mean that we are missing what young people might in fact be both trying to express but also conceal from us and from themselves. Investigating the multiple meanings and contexts in which gender dysphoria emerges may potentially open a vast terrain of previously unacknowledged experience and provide a far-reaching context for understanding the suffering that was previously attributed to being trans. Adopting a transgender identity may be a way to both manage these unformulated feelings and to obscure their origins. Gender-affirming care colludes with this process by erasing the social realities in which childhood gender dysphoria is emerging as an ever-more prevalent phenomenon.

The transgender child as new social discourse

Social theorists argue that discourses imperceptibly become part of the fabric of our social context and determine and delimit how we think about and understand ourselves. Gender theorists posit that discourses about gender determine what is thinkable and available to us in terms of how we can define ourselves as a man or a woman. Judith Butler, one of the most influential gender theorists of our time, has argued that we are created by cultural discourses and there is no self that exists outside of and is not permeated by these discourses. She argues: 'There is only a taking up of the tools where they lie, where the very 'taking up' is enabled by the tool lying there' (Butler, 1990). These tools determine how we behave and live. Nussbaum (1999) explains that, according to Butler, 'by behaving as if there were male and female 'natures', we co-create the social fiction that these natures exist'.

As Brunskell-Evans (2018) has argued, transgenderism, and certainly childhood transgenderism, was not available to us as an idea that could be thought about prior to the recent decades during which these new terms and categories were invented, and associated discourses evolved. These new discourses about trans and transitioning are now very much a part of the fabric of our culture and, by extension, shape the choices we make and the way we live our lives. Trans is a new 'tool', using Butler's language, a new social construction (or 'fiction', in Nussbaum's language), which increasing numbers of children are now picking up, in the same way that previously 'male' and 'female' were utilised as the only available tools which defined what we could become. Trans is now available to our cultural psyche as a new way to re-imagine ourselves, but also as a compelling way to make sense of individual pain and distress. It makes available a discourse that provides an explanation for feelings of alienation and despair when the real issues may not be apparent to the young person, as in the cases of Catherine and Mika. It also provides a pre-fabricated solution that holds out the promise of a new life in a new identity and potentially a new body.

The narrative of the transgender child who can be transformed into a child that expresses their 'true gender self' (Ehrensaft, 2012) through the miracles of modern medicine has firmly established itself in our culture and has an almost mythical and heroic quality. Whilst the new social category of trans promises to 'trouble' gender norms and to help loosen their grip, its concretisation as an identity and biological reality and its appropriation by biomedical psychiatry is problematic. This new social-medical discourse saturates our culture so completely that it now determines how we understand children who are gender non-conforming, and it also shapes how gender non-conforming children understand themselves. Beyond shaping how they make sense of their experience, it conscripts us into an unquestioning acceptance of the necessity of gender-affirming treatment for many children. Further, whilst it promises to liberate children from oppressive gender constraints, we need to consider that the new transgender 'tool' will be utilised by young people in multiple ways, not all of which are necessarily liberating.

In contrast, biomedical psychiatry, with its atomised focus on the individual, understands gender and trans as concrete realities that reside in the individual, rather than new 'tools' that can be appropriated and used. When gender is seen as a kind of bedrock that needs to be accepted and affirmed, it narrows our attention, subverting the possibility of understanding the complexity of how children understand themselves and why they suffer. In the biomedical paradigm, questions about what the new 'tool' of trans is being recruited to do never arise, potentially erasing the real family and social problems constituting the child's distress. As previously proposed, trans can potentially be appropriated by young people in the service of liberation, but it can also be appropriated as a solution to other problems, or to eliminate a part of the self. In particular, it may crystallise as a way for children to make sense of and provide relief from suffering that is generated by the family and social systems in which they dwell.

Conclusion

The biomedical conceptualisation of childhood gender dysphoria erases the multiple contexts in which gender distress arises and collapses space for thinking about the complexity of individual and social experience. It neglects and obscures real relational, family and social issues which may be generating the distress of the child who experiences gender dysphoria. It constrains the range of potential solutions to a singular, concrete pathway consisting of varying combinations of social, medical and surgical transitions. It relegates psychological and family interventions to the status of adjuncts rather than appropriate interventions for a problem that emerges within multiple intersecting systems. Further, if the current wave of gender dysphoria is in fact a communication or protest about the world in which children find themselves, gender-affirming care is an inadequate or even misguided solution. It entails a superficial understanding of the problem and forecloses the possibility of a 'deeper listening'. Such a deeper listening might create a space for imagining more creative and ultimately liberating solutions for gender-distressed children. Seeking to understand what children are expressing beyond its surface content might also require us to seriously confront many problematic social realities, both gender-related and non-gender-related. If we listen and think creatively, perhaps we will find that children's 'engagement with gender, especially when it is transgressive or countercultural, may reveal a creativity and even a politics that can contribute to the erosion (if not destabilization) of the gender system as it presently operates' (Schwartz, 2012). Doing so may make it possible to image a future freer from oppression and in which it is possible for all humans to thrive.

References

Beebe, B., & Lachmann, F. (1998). Co-Constructing Inner and Relational Processes: Self- and Mutual Regulation in Infant Research and Adult Treatment. *Psychoanalytic Psychology*, *15*(4), 480-516.

Bracken, P., Thomas, P., Timimi, S., Asen, E., Behr, G., Beuster, C., Yeomans, D. (2012). Psychiatry beyond the current paradigm. *British Journal of Psychiatry*, *201*(6), 430–434. http://doi.org/10.1192/bjp.bp.112.109447

Brunskell-Evans, H. (2018). Gendered Mis-Intelligence: The Fabrication of 'The Transgender Child'. In H. Brunskell-Evans and M. Moore (Eds.), *Transgender Children and Young People: Born in your own body* (pp.41-63). Newcastle upon Tyne: Cambridge Scholars Publishing.

Butler, J. (1990). *Gender Trouble: Feminism and the subversion of identity*. New York: Routledge.

Capra, F. (1982). *The Turning Point: Science, society and the rising culture*. New York: Simon and Schuster.

Capra, F. and Luisi, P. (2014). *The Systems View of Life: A unifying vision*. New York: Cambridge University Press.

D'Angelo, R. (2018). Psychiatry's ethical involvement in gender-affirming care. *Australasian Psychiatry*, *26*(5), 460-463. http://doi.org/10.1177/1039856218775216

Davies, J.M. and Frawley, M.G. (1992). *Treating the Adult Survivor of Childhood Sexual Abuse. A Psychoanalytic Perspective*. New York: Basic Books.

De Vries A., McGuire J., Steensma T., Wagenaar, E., Doreleijers, T., & Cohen-Kettenis, P. (2014). Young adult psychological outcome after puberty suppression and gender reassignment. *Pediatrics*, *134*, 696-704.

De Vries, A., & Cohen-Kettenis, P. (2012). Clinical management of gender dysphoria in children and adolescents: The Dutch approach. *Journal of Homosexuality*, *59*(3), 301-320. http://doi.org/10.1080/00918369.2012.653300

Department of Health. (2008). *Medical care for gender variant children and young people: answering families' questions*. 1-32. Retrieved from http://webarchive.nationalarchives.gov.uk/20130123193146/http://www.dh.gov.uk/en/Publicationsandstatistics/Publications/PublicationsPolicyAndGuidance/DH_082976.

Ehrensaft, D. (2012). From gender identity disorder to gender identity creativity: True gender self child therapy. *Journal of Homosexuality*, *59*(3), 337-356. http://doi.org/10.1080/00918369.2012.653303
—. (2014). Found in Transition: Our Littlest Transgender People. *Contemporary Psychoanalysis*, *50*(4), 571-592.
Fausto-Sterling, A. (2012). The dynamic development of gender variability. *Journal of Homosexuality*, *59*(3), 398-421. http://doi.org/10.1080/00918369.2012.653310
Fausto-Sterling, A., Crews, D., Sung, J., García-Coll, C., & Seifer, R. (2015). Multimodal sex-related differences in infant and in infant-directed maternal behaviors during months three through twelve of development. *Developmental Psychology*, *51*(10), 1351-1366. http://doi.org/10.1037/dev0000033
Guillamon, A., Junque, C., & Gomez-Gil, E. (2016). A Review of the Status of Brain Structure Research in Transsexualism. *Archives of Sexual Behavior*, *45*(7), 1615-1648. http://doi.org/10.1007/s10508-016-0768-5
Hsieh, S., & Leininger, J. (2014). Resource List: Clinical Care Programs for Gender-Nonconforming Children and Adolescents. *Pediatric Annals*, *43*(6), 238-244. http://doi.org/10.3928/00904481-20140522-11
Jeffreys, S. (2016). *Gender Hurts: A feminist analysis of the politics of transgenderism*. New York: Routledge.
Jennissen, S., Huber, J., Ehrenthal, J. C., Schauenburg, H., & Dinger, U. (2018). Association Between Insight and Outcome of Psychotherapy: Systematic Review and Meta-Analysis. *The American Journal of Psychiatry, 175*(10), 961-969. https://doi.org/10.1176/appi.ajp.2018.17080847.
Jones, M. (2018, February 7). What teenagers are learning from Online Porn. *The New York Times*. Retrieved from https://www.nytimes.com/
Laing R.D. & Esterson, A. (1964). *Sanity, madness and the family*. New York: Basic Books.
Laing, R.D. (1985). Mystification, Confusion and Conflict. In I. Boszormenyi-Nagy and J. Framo (Eds.), *Intensive Family Therapy: Theoretical and practical aspects* (pp.343-363). New York: Routledge.
Lawrence, A. (2014). Treatment of Gender Dysphoria. In G. Gabbard (Ed.), *Gabbard's Treatment of Psychiatric Disorders (Fifth Edition)* (pp.695-719). Arlington: American Psychiatric Publishing.
Littman, L. (2018). Rapid-onset gender dysphoria in adolescents and young adults: A study of parental reports. *PLOS ONE, 13*(8), e0202330. https://doi.org/10.1371/journal.pone.0202330

Murad, M., Elamin, M., Garcia, M., Mullan, R., Murad, A., Erwin, P., & Montori, V. (2010). Hormonal therapy and sex reassignment: A systematic review and meta-analysis of quality of life and psychosocial outcomes. *Clinical Endocrinology*, *72*(2), 214-231. http://doi.org/10.1111/j.1365-2265.2009.03625.x

Nussbaum, M. (1999, February 23). The Professor of Parody. *The New Republic*. Retrieved from https://newrepublic.com/

Olson, K. R., Durwood, L., DeMeules, M., & McLaughlin, K. A. (2016). Mental Health of Transgender Children Who Are Supported in Their Identities. *Pediatrics*, *37*(3), 1-8. http://doi.org/10.1542/peds.2015-3223

Plan International USA. (2018). *The State of Gender Equality for U.S. Adolescents: Survey Highlights*. Retrieved from https://www.planusa.org/docs/state-of-gender-equality-summary-2018.pdf.

Ruti, M. (2017). *The Ethics of Opting Out: Queer theory's defiant subjects*. New York: Columbia University Press.

Schwartz, D. (2012). Listening to children imagining gender: Observing the inflation of an idea. *Journal of Homosexuality*, *59*(3), 460-479. http://doi.org/10.1080/00918369.2012.653314

Seligman, S. (2005). Dynamic systems theories as a metaframework for psychoanalysis. *Psychoanalytic Dialogues*, *15*(2), 285-319. http://doi.org/10.1080/10481881509348832

Singal, J. (2018, July/August). When Children Say They're Trans. *The Atlantic*. Retrieved from https://www.theatlantic.com/

Steensma T., McGuire J., Kreukels B., Beekman A., Cohen-Kettenis P. (2013). Factors associated with desistence and persistence of childhood gender dysphoria: a quantitative follow-up study. *Journal of the American Academy of Child and Adolescent Psychiatry*. 52(6):582–590pmid:23702447

Telfer, M., Tollit, M., Pace, C. and Pang, K. (2018). Australian standards of care and treatment guidelines for transgender and gender diverse children and adolescents. *Medical Journal of Australia*, *209*(3), 132-136. doi: 10.5694/mja17.01044

Thelen, E. (2005). Dynamic Systems Theory and the Complexity of Change. *Psychoanalytic Dialogues*, *15*(2), 255-283. http://doi.org/10.1080/10481881509348831

Violence Policy Centre. (2018). *When Men Murder Women: An analysis of 2015 homicide data*. September 2017. Retrieved from http://www.vpc.org/studies/wmmw2017.pdf

World Economic Forum. (2017). *The Global Gender Gap Report 2017.* Retrieved from http://www3.weforum.org/docs/WEF_GGGR_2017.pdf

World Professional Association for Transgender Health. (2009). *Standards of Care for the Health of Transsexual, Transgender, and Gender Nonconforming People.* Retrieved from https://www.wpath.org/media/cms/Documents/Web Transfer/SOC/Standards of Care V7 - 2011 WPATH.pdf

Zucker, K. (2017). Epidemiology of gender dysphoria and transgender identity. *Sexual Health, 14*(5), 404-411. https://doi.org/10.1071/SH17067

Chapter Five

Gender Development and the Transgendering of Children

Dianna T. Kenny

In August 2017 a teacher from Rocklin Academy Gateway, a primary school in California, USA, conducted a 'transition ceremony' for a so-called transgender five-year-old. The teacher read the class two books—*I am Jazz* and *The red crayon*—that purportedly educate four- to eight-year-old children on transgenderism. Faulty concepts such as 'being born in the wrong body/having a girl's brain in a boy's body' were introduced and discussed, after which a male student left the class dressed as a boy and returned a few minutes later dressed as a girl. The teacher introduced their 'new' classmate to the class, explaining that 'he' was now a 'she' who had a new name and that all 'her' classmates should address 'her' using the pronouns 'she/her' and call 'her' by 'her' new name in future (Laurence, 2017).

The reactions of some of the students were not anticipated by the school. Some of the girls found the ceremony deeply disturbing, asking their parents whether they were going to turn into boys. A male student asked his mother if he, too, could dress as a girl for school because his classmate could do so. Other boys went home to their parents, crying, 'I don't want to turn into a girl' (CBN News, 2017). These reactions deserve careful consideration because they tell us a great deal about gender development in early childhood, a subject that has been sorely neglected in the debate about transgendering young children. Accordingly, in this chapter, I review the research on gender development in normally developing children. I then apply this research to a critique of erroneous assumptions promulgated by trans activists and self-styled paediatric gender therapists. Finally, I discuss how this vast body of literature can be put to good use by professionals dealing with gender dysphoria in young children to assist them to provide sound arguments to parents and teachers as to why young children should not be precipitated into social transition, at least not before their concept of gender is fully established, if at all.

Gender development

Gender development is a complex process, the outcome of which is a complex interplay of genes, gonadal hormones, cognitive, language, and socioemotional development, the child's socialization history, and culture.

Psychologically, the development of gender concepts follows the same stages as those in the development of object and person constancy. Jean Piaget and colleagues (Piaget, 1942, 1947; Piaget & Cook, 1954a, 1954b, 1954c; Piaget & Inhelder, 1969) identified four stages of cognitive development based on the different ways that children of different ages and cognitive capacities construct reality. In the first stage (i.e., sensorimotor stage from birth to ~18 months), the infant interacts with the world via his/her senses and motoric activity. Through repeated experiences of the physical world, the infant gradually develops schemas or organized patterns of behaviour, which gradually cohere into more complex, higher-order schemas. A process of adaptation then occurs that allows information to be assimilated into the new schema or to be accommodated if new knowledge replaces old, less appropriate schemas for the new object. Feedback from the environment determines which schemas are maintained and elaborated and which fade into disuse. In this process, combinations of visual, auditory, tactile, olfactory, and motor representations of objects are combined to form more complex, complete, and permanent representations of objects (and people) in the real world (for a detailed description, see Kenny, 2013).

A critical and central skill acquired at the end of the sensorimotor period is object permanence, defined as the understanding that objects (and people) continue to exist when out of sight. Object permanence is necessary for more complex schema development and for memory. Absence or poorly developed object or person permanence explains the lack of distress in infants in the first few months of life when their mother leaves, and the capacity of other caregivers to attend to the child's needs without protest from the infant. When object permanence is complete, infants understand that their mothers are unique and continue to exist when out of sight, hence their distress when she is absent. Interestingly, babies with secure attachments to their mothers develop person permanence at a younger age than object permanence (Kenny, 2013).

Following Piaget, Kohlberg and Maccoby (1966) proposed that the development of gender constancy (i.e., the notion of the permanence of categorical sex) is a necessary precursor to conformity with culturally defined gender norms; in other words, gender cognitions precede gendered behaviour (i.e., 'I am a boy, therefore I want to behave like a boy'). According to this cognitive-developmental theory, there are three stages in the acquisition of gender constancy: (i) identification of one's own and others' sex, that is, basic gender identity and labelling; (ii) gender stability/ gender permanence (i.e., gender remains stable over time); and (iii) gender constancy (i.e., gender is a fixed characteristic that is not altered by superficial transformations in appearance or activities).

As children develop their concept of gender, they initially focus on the perceptual properties of a person and act as if these properties (e.g., the person's name, long or short hair, pink or blue clothes, etc.) are the defining characteristics of that person. They cannot conserve or retain a person's basic gender identity when outward characteristics change. In other words, they are perceptually bound—they define the concepts of 'male' and 'female' in terms of outward appearances such as hair, clothing, toys, etc. rather than in terms of the person's genitalia or biological sex. Some children older than three years continue to have difficulty conserving sex across perceptual transformations and these difficulties may continue up to the age of seven. This is developmentally normal. Even when preschool children do show gender constancy, it is unlikely that they understand its biological basis, a phenomenon called pseudo-constancy (Eaton & Von Bargen, 1981). Pseudo-constant children provide correct judgements of gender but with incorrect explanations as to their judgement, while children with true constancy offer both correct judgements and explanations. Pseudo-constancy is a transitional process leading to the attainment of true constancy.

Once gender constancy (i.e., consistency and stability of the concept) is achieved, children display lower levels of rigidity or gender stereotypy in gender-based behaviour and become more flexible in their reactions to culturally-derived gender norm violations. This generally occurs around five years of age, although it is interesting to note that children with gender dysphoria are slower to develop gender constancy and engage in less sex-typical behaviour compared with children who can accurately self-label their gender (Zucker et al., 1999). Thus, gender constancy becomes an organizing principle for children's gender beliefs and to some extent, behaviours. Part of the gender development process is the attainment of a sense of the importance of and contentedness with one's gender. Gender typing, a process whereby the child selectively attends to gender cues, same-sex models, same-sex activities, and clothing, is a function of increasing age and emerging constancy (Martin, Ruble, & Szkrybalo, 2002).

The case of Rocklin Academy

Understanding this developmental process helps us to make sense of the distressed reactions of the five-year-old students at Rocklin Primary School. The boys who became distressed had not yet achieved gender constancy, hence their faulty belief that, like the 'transgender' child, they would change sex simply by donning the culturally-defined clothing typically worn by the other sex. It was interesting that some of the girls feared that they would become boys via the same, reverse process, that is, by dressing in boys' clothing. These children demonstrated more advanced cognitive development than the boys because they made a transitive inference; that is, if a boy can change into a girl by donning female clothing, then girls could change into boys by donning boys' clothing. This inference per se is logical, but it distressed these girls because they too had not yet achieved gender constancy, the notion that one's gender is consistent across transformations and stable over time. Research has shown that the age-related reduction in a child's fear of turning into someone of the opposite sex can be attributed to their growing understanding that sex is a fixed and permanent attribute (De Lisi & Gallagher, 1991).

Some of the boys in the class illustrated the phenomenon of social/behavioural contagion, defined as the spread of affect, behaviour, ideas, or attitudes in a group in which one person serves as the model (stimulus) for the imitative actions of others (Christakis & Fowler, 2013). They were perhaps less cognitively developed than the children described above, in that they responded affectively to the care and attention with which their classmate was being treated and wanted to experience that same care and attention. They were therefore focused less on the issue of gender and more on the issue of accruing positive treatment to themselves that they believed resulted from their classmate wearing female clothing.

The response of this last group alerts us to the fact that forces other than cognitive factors are influential in the development of gender concepts. These include social factors. Group processes are critical here. For example, social learning theory argues that children are rewarded for gendered behaviour and this consolidates the cognition, 'I must be a boy/girl' (Mischel, 1966). Parents encourage and reinforce sex-typical play in their children from a young age (Lindsey & Mize, 2001) and a strong association has been found between the amount of encouragement and the level of sex-typical play in normally developing children (Pasterski et al., 2011). By three years of age, children show a marked preference for same-sex playmates and these preferences persist independently of parental involvement, even increasing when adults are not present.

As same-sex imitation increases, cross-sex avoidance heightens at the same time (Grace et al., 2008), possibly because gender constancy has not yet been attained. 'Self-categorization in terms of a social identity maximizes both differences between one's in-group and an out-group and similarities between oneself and other in-group members' (Grace et al., 2008, p.1929). Observing a boy seemingly transforming into a girl would likely disrupt these emergent categorization processes and disturb children's developing cognitions about their gender identity.

Various other psychosocial factors may also play a role in gender identity development, none so powerful as the parental response to cross-gender behaviour in early childhood. Young children cross-dressing as part of play are often viewed as cute and amusing by parents, who may inadvertently reward such behaviour by laughing, taking photos and videos, calling the other parent to view the child, and providing additional 'cross-dress' play clothes in the dressing-up box. Other parental behaviours that undermine or divert healthy gender identity development include parents who treat a child as if s/he were of the opposite sex because they wanted a child of that sex by, for example, refusing to cut the hair of a male child, cross-dressing the child, and supporting the child to engage in activities normally associated with the other sex. Such parental behaviours may interfere with the child's emerging capacity to self-label his/her gender. An example of this process is presented below in the case of Alex.

Finally, the emergence of awareness of sex differences and the display of gender-typical behaviour and preferences occurs primarily within groups after the early gender socialization experiences of infants and toddlers that occurs in families. Developmental Intergroup Theory (Arthur, Bigler, & Ruble, 2009; Liben & Bigler, 2002) is one model that explicates the powerful social psychological processes that operate to create group-based adherence to dimensions of social categorization such as race, religion, politics, sexual orientation, and gender. It appears from research in this area that similar social learning processes operate in learning about gender as those that involve other intergroup or categorization experiences. With respect to gender, mothers' and fathers' behaviours are better predictors of children's gender-role attitudes than parents' gender ideology (Halpern & Perry-Jenkins, 2016).

Thus, the perception of being male or female is both an individual and a group process in that children, in the course of developing a gender identity, assign themselves to a categorical group membership, which asserts new and reinforces existing gender beliefs and attitudes, through processes of modelling and imitation of same-sexed peers and adults. When gendered behaviour is the salient characteristic being observed, children imitate same-sex adults more than same-aged peers, highlighting the importance of same-sex adults as early gender role models:

> On any given day, children are exposed to numerous people exhibiting a variety of behaviors. What they encode as gender normative, however, are those behaviors exhibited most frequently and consistently by multiple members of each gender category. (Grace et al., 2008, p.1929)

These findings suggest that children as a young as three to four years of age can self-categorize along the dimensions of gender and can adapt their behaviour according to the social context in which it occurs; that is, children's imitative behavioural choices are based on their contextually sensitive self-categorizations.

Deconstructing the argument for early social transition

Many self-appointed experts have materialized who travel the world on the conference circuit espousing often scientifically unsupported doctrine that has led to calls for gender affirmation, early social transition, and earlier gender reassignment of putative transgender children. An example is Diane Ehrensaft, a 'paediatric gender therapist.' Her role in life is to liberate gender nonconforming children and youth. Below is a comment she made to KQUD Science:

> We have seen some kids as young as two whose parents are bringing them in because they're beginning to say, 'Me not boy. Me girl.' Social transition can happen as soon as a child has language or the ability to communicate to us who they are. (Ehrensaft, in Brooks, 2018)

Ehrensaft (2016) declared that a female toddler who pulled hair clips out of her hair was communicating a gender message to her parents—i.e., I am a boy (ergo, hairclips are anathema). She then says that another child, born female, who was barely verbal, at around 18 months of age insisted to her parents 'I, boy!' Another child, at one year of age, unsnapped his stud clips on his jumpsuit to, in her words, 'make a dress' which she interpreted as 'preverbal gender communication.'

These comments demonstrate a deplorable misunderstanding of the cognitive capacities and concept formation of preverbal children. Primarily, they assume that these babies have a clear understanding of the concept of gender, that they associate hair clips and dresses with female gender, and that they can recognize and assert their own gender. Further, Ehrensaft communicates inherent assumptions that gender is innate, which discount powerful socialization effects on gender identity, discussed above, if indeed, as Ehrensaft states, babies 'know [their true gender] as early as the beginning of the second year of life; they probably know before, but they are preverbal', in which case we need to be vigilant for preverbal gender communications of the type she described. These assertions stand in stark contrast to research demonstrating that children below the age of three are unlikely to have gender constancy (Ruble et al., 2007) and that gender identification is a very complex interplay between genes, biology (e.g. temperament), sex hormones, social learning, social cognition, psychopathology, and culture.

Further, despite Ehrensaft's (2017) assertions that there are many 'shades of gender' (i.e., transgender, gender fluid, gender hybrid, gender queer, gender smoothie, protogay, prototransgender, gender oreos), there are underlying assumptions of a rigid gender binary in transgender theorizing, which she implies in her statements about babies and their early gender awareness. An example of this confusion is the following: 'Transgender describes persons who do not feel like they fit into a dichotomous sex structure through which they are identified as male or female' (Meier & Labuski, 2013, p.291). For the transgender lobby, the opposite is true. They adhere to rigid and indeed inflated and exaggerated gender binaries, implicitly assuming transgender people to be rigidly binary—male or female—but in the opposite binary to their assigned sex at birth.

As we are seeing throughout this book, there are currently no acceptable theories of gender identity development in children who assert that they are transgender. One study (Olson, Key, & Eaton, 2015) assessed gender identity, gender preference, gender peer preferences, and object preference using both explicit self-report and implicit measures in 32 children aged between 5 and 12 years, who had been living a fully socially transitioned life both at home and in the community. They were recruited through online and in-person support groups and conferences. With such a biased sample, it is perhaps not surprising that results showed that this highly selected group of purportedly transgender children performed similarly to their cisgender siblings and peers in the measures used. This study casts no light on the question of causality as the sample children were already fully inducted into cross-gender identifications, thereby obfuscating a differentiation between social conditioning/social contagion processes such as cross-gender modelling, imitation, and social reinforcement for cross-gender behaviour and innate (cross-)gender awareness. These outcomes support the finding that early social transition and cognitive and/or affective cross-gender identification are associated with the persistence of childhood gender dysphoria into adolescence (Steensma et al., 2013).

Personal and familial psychopathologies in the development of gender dysphoria

> For some [children identifying as 'transgender'], the major issue is cross-gender behaviours or identifications; for others, the gender issues seem to be epiphenomena of psychopathology, exposure to trauma, or attempts to resolve problems such as higher social status or other benefits they perceive to be associated with the other gender. (Drescher & Byne, 2012, p.503)

Recent studies of psychiatric and psychological comorbidities in transgender children presenting to courts or clinics attest to the higher than expected associations between psychopathology and gender dysphoria. For example, in a group of 56 children before the Family Court in Australia, 25 of 39 cases in which family constellation could be discerned lived in single parent families or foster care, with only 14 from two parent families. In this same group of 56 children, 50 percent had a diagnosed psychological disorder. Recent studies have shown a higher prevalence of gender dysphoria in those with Autism Spectrum Disorder (ASD) (De Vries et al., 2010; van der Miesen et al., 2018), major depression (Chen et al., 2016), anxiety and behavioural disorders (Cohen-Kettenis et al., 2003; Zucker et al., 2002), oppositional defiant disorder (ODD), ADHD, (Chen et al., 2016), suicidality/self-harm (Aitken et al., 2016; Peterson et al., 2017), and intellectual disability (Parkes & Hall, 2006).

In a study with a sample of 105 gender dysphoric adolescents and using the Diagnostic Interview Schedule for Children (DISC), anxiety disorders were found in 21 percent of the adolescents, mood disorders in 12.4 percent, and disruptive disorders in 11.4 percent. Males had greater psychopathology compared with females, including comorbid diagnoses (de Vries et al., 2011). Statistics from the Boston Children's Hospital Gender Management Services program revealed a high prevalence of diagnosed psychiatric comorbidities (44 percent), and a history of self-mutilation (21 percent), psychiatric hospitalization (9 percent), and suicide attempts (9 percent) (Spack et al., 2012).

What psychological factors underlie these worrying statistics? To achieve a nuanced understanding, we need to adopt an idiographic approach and examine in detail the individual experiences of young people with gender dysphoria. Below I discuss four pertinent case examples that shed light on the complex dynamics underpinning gender dysphoria.

Case examples

Alex (a biological female), aged 12, petitioned the Family Court of Australia to permit her to transition. The Court made orders allowing the commencement of puberty-suppressing hormone medication because of the intense distress Alex felt at her emergent feminine body. At 17, the Court granted permission for a double mastectomy. Psychiatric evidence indicated a traumatic childhood, in which Alex's mother rejected her completely. However, she had a close and idealized relationship with her father, who wanted her to be a boy and who treated her as such, even teaching her to urinate in the standing position. He died suddenly when Alex was six. Psychiatric evaluation revealed significant early trauma and concluded that 'Alex's cross-gender identification appears to have emerged in the context of an idealised, physically close relationship with her father, rejection and abandonment by her mother, and her father's desire for her to be a male ... Her investment as male simultaneously expresses anger towards her mother and maintains closeness with her dead father... in the context of her incomplete mourning for him' (Kissane, 2009).

Saketopoulou (2015) asks the question: 'Is the patient gender-confused, failing at adequate reality testing, or is the environment unable to mirror an internal experience that doesn't meet expectable forms of gender?' (p.774). In this case, this child's gender dysphoria formed part of a complex psychopathology that had its origins in early attachment ruptures that resulted in non-identification with an abandoning, rejecting mother and over-identification with a father who was unable to value and accept his child as a daughter, but who rewarded the child with acceptance and approval for trying to be a son. The sudden death of Alex's father at the age of six left the child bereft; becoming male as opposed to identifying as male became the only way this child could maintain contact with the father in fantasy, by becoming the father, and to find self-acceptance, since she had been rejected by both parents as a female child.

Ariel, a transfemale, aged 13, who had commenced puberty blockers, insisted on being called by the name of a different Disney princess every day, until she settled on the name, Ariel:

> 'I remember… when everyone was talking about having babies and it really makes me upset. I don't want to tell them to stop talking about it… but it hurts my feelings when they're talking about it… I am like a girl, but can I have the pain of labour? For a lot of people, it is hard for them to understand, but I don't want to burden them with that. Sometimes I just walk away and sometimes I try to get into the conversation, but it's hard'. Her remarkably perceptive friend then says, 'You can get so close to being a girl but you can't get to that exact point. Is that what upsets you?' Ariel says 'Yeah, that's exactly how I feel, the thing with having a baby, I can never be fully there. It is a natural thing that happens. I buy a bra but it's not to hold in my boobs—it is an illusion. It felt like an act, so I feel lost sometimes.' (https://www.youtube.com/watch?v=sTfQ44HFu6k)

Ariel articulates her lived experience of impersonating a girl rather than becoming one or being one. None of the culturally feminine ideals and products with which she surrounds herself can fully convince her that she is female. She acknowledges that it is an 'illusion', 'an act', and she feels 'lost' that a true gender identity eludes her.

Jesper, a transmale, aged 13, had this to say about the role of the internet in his 'coming out as trans':

> The internet is the best place for trans people, it is the best place you can go to if you are scared about talking to anyone. TUMBLR Oh, My God! TUMBLR! YouTube too. That's how I found out that I was trans—it was from a YouTube video… (video deleted, but see uppercase CHASE1, 2017 as an example)

This young person appeared to have no caring, empathic adult with whom to share his identity/gender confusion and turned to the internet to seek out like minds, that is, to find his 'true' in-group. Seeking and finding membership in a valued in-group enhances self-esteem and feelings of belonging and affiliation (Buck et al., 2013). Feeling alienated and marginalized in the 'real' world (he was 'scared about talking to anyone'), the virtual world of the internet appeared to provide a substitute community missing in the child's real world. However, there is no opportunity to reality-test in such a process, and this young person may have commenced down a dangerous path in order to attach himself to an in-group, where he could experience social inclusion. One can also characterize this process as social contagion, since it is likely that the transgender in-group to which he became attached comprised members who were also seeking inclusion and validation in the ether because the real world had rejected or marginalized them. For another example of this process, see Olszewski (2016), in which a young boy says that the internet is 'hugely important' particularly when parents are disapproving:

> For as long as I can remember, I always felt male. I did come out to my parents as lesbian, sometime around seventh [grade]. I thought, 'Oh well, I seem to wear boys' clothes all the time, I feel masculine, and I realise that I like girls, so then I thought, 'OK, I must be a lesbian. That was tough. My dad, he wouldn't have any part of it. He said, 'This is not a world that you are going to be a part of.' Then, when I got to my freshman year, I identified as trans, so I came out to them again as a transmale. I always had a hard time making friends. I was a very strange kid. I would just feel bad because every day I went to school, I felt like everybody wanted me to go; nobody wanted me there. One of the girls said, 'Man, you are an ugly dyke. You are a lesbian.' I went from shaky, to unstable, to almost impossible. I started drifting off to a very violent place in my head. I had thoughts of harming my family. It got so bad, I felt like a threat to my family, and to myself. One night, I went down to my mom and said that I wanted her to take me to a hospital; I wanted to get locked up. (John, age 16, transmale)

This transcript demonstrates the confusion experienced by some young people with gender dysphoria as to their sexual orientation and gender identity, with some believing they are transgender when they are in fact homosexual/lesbian. Existing theories of transgender also conflate these two dimensions, based as they are on a 'coming out' model developed for people with lesbian/gay orientations. There has also been a tendency to conflate gender identity with sexual orientation in seeking causal explanations (Katz-Wise et al., 2017). Although I have no further detail on this young person, it is clear that he was experiencing significant psychiatric morbidity, which is a common feature of presentations of children with gender dysphoria, as discussed earlier.

Conclusions

Gender development is the product of genetic, biological, cognitive, and social factors. These factors are not monoliths acting independently. Rather, they form a complex set of interactions that influence gender development, perhaps in different ways for different children. In this chapter, I have addressed the multi-faceted process of normal gender development in young children, highlighting the need for those working with purported transgender children to have a sophisticated and nuanced understanding of this process in order to prevent a precipitous social transition from which a child may never recover. Detailed examinations of the child's life and primary caregiver and peer relationships need to be undertaken to rule out gender dysphoria secondary to trauma, psychopathology, such as ASD, ADHD, or intellectual disability, issues in the parent-child relationship, or social contagion.

References

Aitken, M., VanderLaan, D. P., Wasserman, L., Stojanovski, S., & Zucker, K. J. (2016). Self-harm and suicidality in children referred for gender dysphoria. *Journal of the American Academy of Child & Adolescent Psychiatry, 55*(6), 513-520. doi: 10.1016/j.jaac.2016.04.001

Arthur, A. E., Bigler, R. S., & Ruble, D. N. (2009). An experimental test of the effects of gender constancy on sex typing. *Journal of Experimental Child Psychology, 104*(4), 427-446. doi: https://doi.org/10.1016/j.jecp.2009.08.002

Brooks, J. (2018). *Is three too young for children to know they're a different gender? Transgender researchers disagree.* Retrieved from https://www.kqed.org/futureofyou/440851/can-you-really-know-that-a-3-year-old-is-transgender

Buck, D. M., Plant, E. A., Ratcliff, J., Zielaskowski, K., & Boerner, P. (2013). Concern over the misidentification of sexual orientation: Social contagion and the avoidance of sexual minorities. *Journal of Personality and Social Psychology, 105*(6), 941-960.

CBN News. (2017). Kindergarteners scared they will be 'turned into boys' after school's transgender celebration. *CBN News.* Retrieved from http://www1.cbn.com/

Chen, M., Fuqua, J., & Eugster, E. A. (2016). Characteristics of referrals for gender dysphoria over a 13-year period. *Journal of Adolescent Health, 58*(3), 369-371. doi: https://doi.org/10.1016/j.jadohealth.2015.11.010

Christakis, N. A., & Fowler, J. H. (2013). Social contagion theory: examining dynamic social networks and human behavior. *Statistics In Medicine, 32*(4), 556-577.

Cohen-Kettenis, P. T., Owen, A., Kaijser, V. G., Bradley, S. J., & Zucker, K. J. (2003). Demographic characteristics, social competence, and behavior problems in children with gender identity disorder: A cross-national, cross-clinic comparative analysis. *Journal of Abnormal Child Psychology, 31*(1), 41-53. doi: 10.1023/a:1021769215342

De Vries, A. L., Noens, I. L., Cohen-Kettenis, P. T., van Berckelaer-Onnes, I. A., & Doreleijers, T. A. (2010). Autism spectrum disorders in gender dysphoric children and adolescents. *Journal of Autism and Developmental Disorders, 40*(8), 930-936.

De Lisi, R., & Gallagher, A. M. (1991). Understanding of gender stability and constancy in Argentinean children. *Merrill-Palmer Quarterly*, 37(3), 483-502.

Drescher, J., & Byne, W. (2012). Gender dysphoric/gender variant (gender dysphoria/gv) children and adolescents: Summarizing what we know and what we have yet to learn. *Journal of Homosexuality, 59*(3), 501-510. doi: 10.1080/00918369.2012.653317

Eaton, W. O., & Von Bargen, D. (1981). Asynchronous development of gender understanding in preschool children. *Child Development, 52*, 1020-1027.

Ehrensaft, D. (2016). *Barrettes* [Video file]. Retrieved from http://www.vimeo.com/185149379

—. (2017). *Gender spectrum* [Video file]. Retrieved from http://www.youtube.com/watch?v=HpE3d69SiDU

Grace, D. M., David, B. J., & Ryan, M. K. (2008). Investigating preschoolers' categorical thinking about gender through imitation, attention, and the use of self-categories. *Child Development, 79*(6), 1928-1941.

Halpern, H. P., & Perry-Jenkins, M. (2016). Parents' gender ideology and gendered behavior as predictors of children's gender-role attitudes: A longitudinal exploration. *Sex Roles, 74*(11-12), 527-542.

https://www.youtube.com/watch?v=sTfQ44HFu6k [Video file]. Retrieved 21 May, 2018, from https://www.youtube.com/watch?v=sTfQ44HFu6k

Katz-Wise, S. L., Budge, S. L., Fugate, E., Flanagan, K., Touloumtzis, C., Rood, B., Perez-Brumer, A., & Leibowitz, S. (2017). Transactional pathways of transgender identity development in transgender and gender-nonconforming youth and caregiver perspectives from the Trans Youth Family Study. *International Journal of Transgenderism, 18*(3), 243-263.

Kenny, D. T. (2013). *Bringing up baby: the psychoanalytic infant comes of age*. London: Karnac.

Kissane, K. (2009). *Young people, big decisions*. Retrieved from https://www.smh.com.au/national/young-people-big-decisions-20090504-arxc.html

Kohlberg, L., & Maccoby, E. (1966). *The development of sex differences*. Stanford, CA: Stanford University Press.

Laurence, L. (2017). *Kindergarten celebrates 5-year-old transgender 'transition;' kids traumatized.* Retrieved from https://www.lifesitenews.com/news/kindergarten-celebrates-5-year-old-transgender-transition-kids-traumatized

Liben, L. S., & Bigler, R. S. (2002). The developmental course of gender differentiation: Conceptualizing, measuring, and evaluating constructs and pathways. *Monographs of the Society for Research in Child Development, 67*(2), vii-147.

Lindsey, E. W., & Mize, J. (2001). Contextual differences in parent–child play: Implications for children's gender role development. *Sex Roles, 44*(3-4), 155-176.

Martin, C. L., Ruble, D. N., & Szkrybalo, J. (2002). Cognitive theories of early gender development. *Psychological Bulletin, 128*(6), 903-933.

Meier, S. C., & Labuski, C. M. (2013). The demographics of the transgender population. In A. K. Baumle (Ed.), *International handbook on the demography of sexuality* (pp.289-327). New York: Springer.

Mischel, W. (1966). A social learning view of sex differences in behavior. In E. Maccoby (Ed.), *The development of sex differences* (pp.57-81). Stanford, CA: Stanford University Press.

Olson, K. R., Key, A. C., & Eaton, N. R. (2015). Gender cognition in transgender children. *Psychological Science, 26*(4), 467-474. doi: 10.1177/0956797614568156

Olszewski, T. (2016, June 6). *Transgender in the Media and Internet* [Video file]. Retrieved from https://www.youtube.com/watch?v=eYOuqgoxAik

Parkes, G., & Hall, I. (2006). Gender dysphoria and cross-dressing in people with intellectual disability: a literature review. *Mental Retardation, 44*(4), 260-271.

Pasterski, V. L., Geffner, M. E., Brain, C., Hindmarsh, P., Brook, C., & Hines, M. (2011). Prenatal hormones and childhood sex segregation: Playmate and play style preferences in girls with congenital adrenal hyperplasia. *Hormones and Behavior, 59*(4), 549-555. doi: https://doi.org/10.1016/j.yhbeh.2011.02.007

Peterson, C. M., Matthews, A., Copps☐Smith, E., & Conard, L. A. (2017). Suicidality, self☐harm, and body dissatisfaction in transgender adolescents and emerging adults with gender dysphoria. *Suicide and Life-Threatening Behavior, 47*(4), 475-482.

Piaget, J. (1942). The three fundamental structures of psychic life: rhythm, equilibrium, and grouping. *Psychologie, 1*, 9-21.

—. (1947). *The psychology of intelligence.* Oxford, UK: Armand Colin.

Piaget, J., & Cook, M. (1954a). *The construction of reality in the child.* New York: Basic Books.
—. (1954b). The development of causality. In J. Piaget & M. Cook (Eds.), *The construction of reality in the child* (pp.219-319). New York: Basic Books.
—. (1954c). The development of object concept. In J. Piaget & M. Cook (Eds.), *The construction of reality in the child* (pp.3-96). New York: Basic Books.
Piaget, J., & Inhelder, B. (1969). *The psychology of the child.* London: Routledge and Kegan Paul.
Ruble, D. N., Taylor, L. J., Cyphers, L., Greulich, F. K., Lurye, L. E., & Shrout, P. E. (2007). The role of gender constancy in early gender development. *Child Development, 78*(4), 1121-1136.
Saketopoulou, A. (2015). Diaspora, exile, colonization: Masculinity dislocated. *Studies in Gender and Sexuality, 16*(4), 278-284.
Spack, N. P., Edwards-Leeper, L., Feldman, H. A., Leibowitz, S., Mandel, F., Diamond, D. A., & Vance, S. R. (2012). Children and adolescents with gender identity disorder referred to a pediatric medical center. *Pediatrics, 129*(3), 418-425. doi: 10.1542/peds.2011-0907
Steensma, T. D., McGuire, J. K., Kreukels, B. P. C., Beekman, A. J., & Cohen-Kettenis, P. T. (2013). Factors associated with desistence and persistence of childhood gender dysphoria: A quantitative follow-up study. *Journal of the American Academy of Child & Adolescent Psychiatry, 52*(6), 582-590. doi: https://doi.org/10.1016/j.jaac.2013.03.016
UppercaseCHASE1. (2017, June 21). *FEAR OF TRANSITIONING* [Video file]. Retrieved from https://www.youtube.com/watch?v=TfpPldcFfUQ
Van der Miesen, A. I. R., Hurley, H., Bal, A. M., & de Vries, A. L. C. (2018). Prevalence of the wish to be of the opposite gender in adolescents and adults with autism spectrum disorder. *Archives of Sexual Behavior, 47*(8), 2307-2317. doi: 10.1007/s10508-018-1218-3
Zucker, K. J., Bradley, S. J., Kuksis, M., Pecore, K., Birkenfeld-Adams, A., Doering, R. W., Mitchell, J. N., & Wild, J. (1999). Gender constancy judgments in children with gender identity disorder: Evidence for a developmental lag. *Archives of Sexual Behavior, 28*(6), 475-502. doi: http://dx.doi.org/10.1023/A:1018713115866
Zucker, K. J., Owen, A., Bradley, S. J., & Ameeriar, L. (2002). Gender-dysphoric children and adolescents: A comparative analysis of demographic characteristics and behavioral problems. *Clinical Child Psychology and Psychiatry, 7*(3), 398-411. doi: 10.1177/1359104502007003007

Chapter Six

Sex Development: Beyond Binaries, Beyond Spectrums

Nathan Hodson

The 'orthodox' transgender idea is that sex is socially constructed. This proposition relies upon the claim that intersex people disrupt the binaristic view of biological sex. In this chapter I argue that understanding and differentiating the experiences of trans people and intersex people is fundamental to thinking coherently about the categories of transgenderism and intersex. I outline who the category of people defined as intersex includes, and what intersex means from the contrasting positions of those who identify as intersex and those who identify as transgender. I demonstrate the existence of intersex conditions does not call into question the existence of natural, biological sex difference. Rather, the diversity and complexity of human sex development reveals a set of patterns of development of intersex conditions which incorporate or combine elements from the two sexes, male and female, and entails recognition of the relative infrequency of, and special challenges faced by, intersex people.

Meanings of intersex and transgender

Over the last sixty years or so, meanings of 'intersex' and 'transgender' have evolved through a complex process of linkage and disambiguation. In 1996, the psychologist John Money explained that working on intersex had allowed him to formulate a new and highly influential theory of gender:

> The majority of people who contributed to this new meaning of gender were hermaphrodites or intersexes. To them social science and social history overall owe a debt of gratitude. It is impossible to write about the

social and political history of the second half of the twentieth century without reference to the concept of gender.

Working on intersex patients in the 1950s, Money had felt the need for a term to describe a patient's sense of their sexed identity. At first, he used 'gender role' and then borrowed 'gender identity' from the psychoanalyst Robert Stoller. Money (1996) explains that:

> Gender role is what you say and do from which other people piece together their own version of your gender identity. Your gender identity is more inclusive than your gender role: it includes ideation, imagery, and unspoken text that may be known to you alone.

Money extended his thinking about the formation of gender identity in intersex patients to those born unambiguously male or female. In doing so, he helped produce the fascination with gender which gripped the latter part of the twentieth century. It became possible to conceptualise 'gender identity' as dislocated from biological sex when new medical technologies for the first time made it possible for doctors to change the bodies of those born with indeterminate genitals and to assign them to a sex. As Bernice Hausman (1995) argues:

> Clinicians developing protocols for the treatment of intersexuality needed a theory to help them decide what sex to place their subjects in. After all, for the first time in history, doctors could 'fix' intersexuality with the relatively new sciences of endocrinology and plastic surgery.

Alice Dreger (2015) points out that at the end of the twentieth century, some intersex people who had been subject to surgery as babies or children began to campaign against interventions which left many with damaged sexual sensation, incontinence or repetitive infections. In contrast to previous medical practice, where parents had been encouraged to accept early medical intervention on babies born with intersex conditions and to accept the maximum number of medical interventions thereafter, intersex activists argued that doctors should only intervene if strictly medically necessary. According to Dreger, intersex activists aim to allow people to inhabit the differently configured bodies in which they were born until they are old enough to make any decision themselves.

The medical interventions challenged by intersex activists were, by contrast, fundamental to the twentieth century invention of the transsexual. There are a number of case studies of early transsexuals who claimed to be intersex in order to circumvent legal and ethical constraints which disallowed medical intervention on healthy bodies. The plastic surgeon Sir

Harold Gillies used the fiction of an intersex condition to justify the world's first phalloplasty for Michael (born Laura Maud) Dillon in the late 1940s. He said: 'I will put you down as an acute hypospadias' (Hodgkinson, 2015, p.84). In the case of Agnes, it was the patient who fooled the doctors. The nineteen-year-old had been 'living undetected as a young woman for about two years' and appeared to present as 'a unique type of a most rare disorder'. Robert Stoller (2015, p.133) believed that Agnes had:

> Testicular feminization syndrome, a condition in which the testes are producing oestrogens in sufficient amount that the genetically male foetus fails to be masculinized and so develops female genitalia. This particular case was unique in that the patient was completely feminized in her secondary sex characteristics (breasts and other subcutaneous fat distribution; absence of body, facial, and limb hair; feminisation of the pelvic girdle; and very feminine and soft skin) with a nonetheless normal-sized penis and testes.

Although Agnes obtained genital surgery on the basis of her (sic) unique intersex condition, five years later it was revealed that 'she had never had a biological defect that had feminized her, but that she had been taking oestrogens (stolen from her mother) since age twelve' (Stoller, 2015, p.135). Stoller was embarrassed by what he saw as his mistake, but the idea of intersex has regularly been used by transsexuals in an attempt to offer a biological explanation for their gender identity (Hausman, 1995).

An interesting phenomenon has arisen with regard to political activism between the two lobby groups. On the one hand, contemporary transgender lobby groups such as Mermaids argue passionately for a 'maximum' approach to gender dysphoria in children and use the danger of teen suicide to argue for the earliest possible medical intervention in the form of puberty blockers, followed by cross-sex hormones and genital surgery. On the other hand, lobby groups such as IntersexUK, which campaigns to end stigma around intersex variations, argue passionately against a 'maximum' approach to disorders of sexual development. In contrast to transgender activists, the demands of intersex campaign groups are to put an end to mutilating and 'normalising' practices such as genital surgeries. Where transgender lobby groups insist upon medicalisation for children and campaign for the age of consent for hormone therapy and surgery to be lowered, intersex lobby groups campaign for surgery on children to be outlawed (Kleeman, 2016).

Sex Development: Beyond Binaries, Beyond Spectrums 111

Are intersex people trans?

There are different perspectives within the transgender affirmative associations and communities about whether intersex people are transgender. Figure 6-1 shows a trans umbrella schematic produced by the National Academic Advising Association (NACADA) based in the US which situates intersex underneath the trans umbrella (National Academic Advising Association, 2014).

The Trans Umbrella

Important Note: These words are social constructs developed over time. New language is constantly formed to unite community members as well as divide groups by experience, politics, and other group memberships. I use the word "Trans" to serve the purpose of inclusion for all listed below, allies, partners, and families.

Transgender
An "umbrella term" for someone whose self-identification, anatomy, appearance, manner, expression, behavior and/or other's perceptions of challenges traditional societal expectations of congruent gender expression and designated birth sex.

Transsexual
Individuals whose designated sex at birth does not match their personal sex/body identity and who, through sex reassignment surgery and hormone treatments, may seek to change their physical body to match their gender identity. Transsexuals can be male-to-female (MTF) or female-to-male (FTM). Transsexuals' sexual identification can be heterosexual, gay, lesbian, bisexual, etc.

Crossdressers
People, often heterosexual men, who are comfortable with their birth assigned gender and will *privately dress* or take on the mannerisms of the "opposite" gender for personal gratification.

Drag Performers
People who dress and *theatrically perform* like the "opposite" gender for entertainment, play, expression, or eroticism. Males are referred to as Drag Queens and females are referred to as Drag Kings. Some identify as trans and others do not.

Intersex Condition
"Intersex is a socially constructed category that reflects real biological variation in reproductive, sexual, or hormonal anatomy. Though usually thought of as an inborn condition, intersex anatomy doesn't always show up at birth.

Gender Variant/Queer
People who find other gender categories constraining. Their gender identities and/or expression is consciously not consistent with conventional standards for masculine or feminine behavior or appearance. Some identify as a blend, as androgynous, or as neither gender.

Figure 6-1: The Trans Umbrella from NACADA includes intersex conditions.

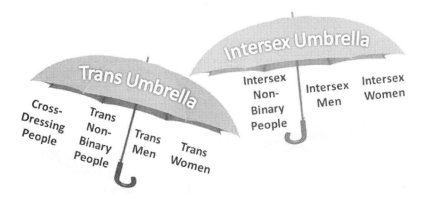

Figure 6-2: The Trans Umbrella and the Intersex Umbrella are different according to Scottish Trans.

Figure 6-2, produced by Scottish Trans, is careful to distinguish trans from intersex (Scottish Trans, 2015).

Intersex Human Rights Australia (2011) also draws a distinction between intersex and trans:

> Intersex is not a part of the trans umbrella (such as transgender or transsexual) nor is intersex a form of gender diversity, because intersex is not about gender, or transition. Intersex is about bodies; about congenital physical differences in sex characteristics.

Intersex Human Rights Australia's objection to the inclusion of intersex under the trans umbrella stems from the conviction that it is crucial to distinguish intersex politics from transgender ideology:

> Intersex, transgender, and same sex attraction are distinct concepts and issues, and people with intersex variations face distinct health and human rights issues. Intersex and transgender are fundamentally different categories, although those with disorders of sexual development may, in some cases, experience problems with gender identity. Indeed, intersex activism has developed in ways that are antithetical to transactivism.

Defining intersex

The Office of the High Commissioner for Human Rights (OHCHR) defines intersex in terms of the typical definitions for male or female bodies:

> Intersex people are born with physical or biological sex characteristics (such as sexual anatomy, reproductive organs, hormonal patterns and/or chromosomal patterns) that do not fit the typical definitions for male or female bodies. For some intersex people these traits are apparent at birth, while for others they emerge later in life, often at puberty. (OHCHR, 2016)

According to the OHCHR, many of the challenges facing intersex people emerge from discrimination and oppression, which arise because people find it difficult to know whether to treat them as male or female. Because of this, intersex people, especially children, remain under threat of harmful and unnecessary medical intervention to fit them into a recognisable sex category:

> ...intersex infants, children and adolescents are subjected to medically unnecessary surgeries, hormonal treatments and other procedures in an attempt to forcibly change their appearance to be in line with societal expectations about female and male bodies.

The Darlington Statement (2017) is a consensus from several intersex groups in Australia and New Zealand which defines intersex against 'stereotypes' and is deliberately inclusive and open, fundamentally focussed on respect for the autonomy of intersex people. The Intersex Society of North America (2011) reiterated the diversity of bodies among intersex people, saying: '"Intersex" is a general term used for a variety of conditions in which a person is born with a reproductive or sexual anatomy that doesn't seem to fit the typical definitions of female or male'. What matters, according to the Intersex Society of North America, is not the specifics of the definition of intersex, but protection from oppressive and disabling attitudes which promote unconsented genital surgery:

> Rather than trying to play a semantic game that never ends, we at ISNA take a pragmatic approach to the question of who counts as intersex. We work to build a world free of shame, secrecy, and unwanted genital surgeries for anyone born with what someone believes to be non-standard sexual anatomy.

The United Nations Special Rapporteur (2013) criticises genital surgery to 'normalise' children born with ambiguous genitalia, highlighting the risk of infertility as well as sexual side-effects:

> These procedures [so-called genital-normalising surgeries] are rarely medically necessary, can cause scarring, loss of sexual sensation, pain, incontinence and lifelong depression and have also been criticized as being unscientific, potentially harmful and contributing to stigma.

Children who are born with atypical sex characteristics are often subject to irreversible sex assignment, involuntary sterilisation, and involuntary genital-normalising surgery, performed without their informed consent or that of their parents, 'in an attempt to 'fix' their sex', leaving them with permanent, irreversible infertility and causing severe mental suffering.

Pragmatic rights-focussed definitions characterise intersex as primarily the condition of people whose sex organ anatomy puts them at risk of non-consensual genital surgery to help them fit into the 'social norm'. The use of distinct definitions protects the integrity of the intersex movement which coheres around the need to protect children and adults with objectively verifiable, innate anatomical variations.

A new organisation, the Accord Alliance (2019), has replaced the Intersex Society of North America and now terms intersex conditions 'Disorders of Sex Development' (DSD), with the intention of providing an objective, systematic, nosological classification. However, this terminology has not proved entirely uncontroversial. A political dimension of the concept 'intersex' makes it a powerful tool to fight oppression, but, at the same time, it means that the term can be a heavy burden for parents whose child is born with atypical genitalia.

Elizabeth Reis (2007) argues that the word 'disorder' is inaccurate for describing intersex conditions and suggests 'divergence' might work better. However, in some intersex conditions, where there are broader endocrine problems in particular infertility, she argues the pathological dimension should be acknowledged. Feder and Karkazis (2008) echo this sentiment. They suggest that generalising about the intersex community is problematic, saying 'a focus on the specific disorders in question ... holds immediate promise for de-medicalizing aspects of the condition that have been improperly pathologized':

> If DSD merely replaces intersex, then it serves only to reinforce a history of medicalization that has brought much physical and emotional pain. For those who have refused to identify as intersex, however, if 'DSD' promotes focus on the medical issues associated with intersex conditions, displacing concerns with gender identity, then it may bring welcome clarification. For those who have refused to identify as intersex, however, the term DSD brings a welcome clarification that theirs is a medical condition, not an identity.

Disorders of sex development

At the beginning of the twentieth century, some doctors believed they could identify 'true' sex in the indeterminate cases of babies born with intersex conditions, based on a microscopic analysis of the gonad (Griffiths 2018a, 2018b). This seemed the simplest approach, albeit one lacking sensitivity to people raised in keeping with their external genitalia who later discover gonads out of keeping with their apparent sex (males with ovaries, females with testes). But as an approach, it is ultimately confounded by the very rare ovo-testicular disorders of sex development, sometimes called mixed gonadal dysgenesis or, previously, 'true hermaphroditism', where somebody's gonads are neither both testicles nor both ovaries.

Ovo-testicular disorders of sex development are the archetypal conditions of atypical sex development and those most closely associated with the now outmoded term hermaphroditism (National Organization for Rare Disorders, 2018). These conditions are characterised by atypical gonadal development, neither a pair of ovaries nor a pair of testes, but some other combination of ovaries, testes and ovotestes (gonads made of a combination of ovarian and testicular tissue). This leads to atypical internal and external sex organs. One cause of ovo-testicular disorder is chromosomal mosaicism. Most people have 46 chromosomes in 23 pairs including either XX chromosomes or XY chromosomes in every cell in their body. In general, people have the same cells. But, due to very early development, a tiny number of people have a body with different sets of chromosomes in each cell and thus more than one karyotype. Depending on the proportion of the cells and the organs, this situation might manifest differently. Mosaicism can cause mixed gonadal dysgenesis when some cells contain a Y chromosome and some cells contain only X chromosomes. This may be called 46XX/46XY when the two lines are XX and XY. It could also be, for example, 46XX/47XXY when the mosaic has cells with an extra sex chromosome. The prevalence of ovo-testicular disorders is thought to be less than one in 20,000 (National Organization for Rare Disorders, 2018).

Congenital Adrenal Hyperplasia (CAH) comprises the majority of the purported frequency of DSD and can affect both males and females. Females with classical CAH are born with an enlarged clitoris which may include a central urethra (Witchel, 2018). Although the majority of paediatric gynaecologists have supported early surgery to make the XX girl's external appearance less male-like with cliteroplasty, vaginoplasty, and labiaplasty during the first two years of life, surgical intervention is cosmetic. Surgery does not make the child more female than she actually is and is not without problematic side-effects, such as urinary incontinence and vaginal stenosis (Merke & Poppas, 2014).

Interestingly, the bodies of children identified as transgender are phenotypically normal, unambiguously either male or female and perfectly healthy. This raises serious political questions about the ethics of medically transitioning, which it is beyond the scope of this chapter to explore, but the required analysis is found elsewhere in this volume. Neither ovo-testicular disorder nor congenital adrenal hyperplasia demonstrate that sex development is regularly socially constructed. What these conditions demonstrate is that a small percentage of people face uniquely difficult decisions regarding medical interventions upon their sexual development, decisions from which the rest of the population are thankfully spared.

Is sex a social construct?

The concepts that sex is a social construct or a spectrum are best articulated by Anne Fausto-Sterling (2018) and Amanda Montañez (2017), although their discussions both entail significant omissions.

Fausto-Sterling (2018) objects to a 'binaristic' view of sex, arguing that sex is a bimodal continuum. She initially (1993) proposed three sexes need to be added to the familiar categories 'male' and 'female': 'herms' are those people with ovo-testicular disorders of sex development; 'merms' are those people who have testes and some female external genitalia; while 'ferms' are those people with male external sex organs and ovaries. These last two groups include, respectively, people with Complete Androgen Insensitivity Syndrome and classical Congenital Adrenal Hyperplasia, among other conditions.

However, these groups are still defined by Fausto-Sterling in terms of the traditional two sexes and sexual dimorphism: each definition is contingent upon the male-female, sperm-egg, and ovary-testis dichotomies. They combine elements of the sexes 'male' and 'female' rather than introduce a new sex chromosome, gonad or gamete to support the discovery of a completely new sex. Throughout biology, sex is understood in terms of reproduction, but there is no suggestion that Fausto-Sterling's additional categories provide a better explanation for human reproduction than the two-sex model.

Fausto-Sterling (2000) subsequently claimed that she was writing with her 'tongue firmly in cheek' and did not believe in the existence of five sexes. Sadly, any retraction was lost in the wider discourse, leading to a popular myth that science has demonstrated the existence of five sexes (Koyama, 2006), a myth which continues to be cited uncritically in newspaper articles more than two decades later (Heggie, 2015).

Blackless *et al* (2000) estimated that 1.5-2% of people deviate from a typical male or female body, which ultimately led to a misconception that this proportion of the population is intersex. Blackless et al (2000) estimated that 1.5-2% of people deviate from a typical male or female body, which ultimately led to a popular misconception that this proportion of the population is intersex. A recent article claims 'being intersex is almost as common as having red hair, yet no one ever talks about it' (Delisle, 2018).

Montañez (2017) provides a more sophisticated analysis than Fausto-Sterling, arguing that the spectrum of sex is set out across multiple layers: chromosomes, genes, hormones, internal and external sex organs, and the secondary sex characteristics that develop at puberty. She maps various routes through development and labels them as typical biological female and typical biological male, with all intersex conditions in between. Her work shows that intersex conditions, on the whole, are discrete from one another: there are set patterns which repeat themselves. Klinefelter syndrome, for example, is a repeated and recognisable pattern of (usually) two X chromosomes and a Y chromosome; androgen insensitivity is a known mechanism which has a predictable presentation; and mosaic ovo-testicular disorders follow paths. These repeating patterns show that sex development, whether typical or atypical, manifests in discrete groups rather than along a sliding scale, and this makes it nonsensical to call sex a spectrum. Her model also confines typical female and male development patterns to the edges of the diagram, wrongly creating the impression that people are spread evenly along the spectrum; in fact over 98% of people experience typical female or male development.

Sax argues that misrepresenting the frequency of atypical sex development does a disservice to intersex people:

> If the term intersex is to retain any clinical meaning, the use of this term should be restricted to those conditions in which chromosomal sex is inconsistent with phenotypic sex, or in which the phenotype is not classifiable as either male or female. The birth of an intersex child, far from being 'a fairly common phenomenon' is actually a rare event, occurring in fewer than 2 out of every 10,000 births. (Sax, 2002)

All human sex development draws on male and female patterns – there is no third gamete or third gonad. Most children are consistently male or consistently female, but there are exceptions which complicate a binary view. But the spectrum approach also misrepresents the biology. Moving beyond flawed simplifications means acknowledging that most people have unambiguous biological sex but that active (human right-based) choices must be made for an important minority of children. We have seen that understanding and differentiating between the experiences of transgender people and intersex people, and the different medical models applied to their bodies in the name of progressivism, throws light on how to think about the situations of trans and intersex children.

Conclusion

Categorising every child as either simply male or simply female would involve difficult decisions particularly with respect to possible medical intervention. But attempts to conceptualize sex as a spectrum have overlooked the 99% of the population with no ambiguity regarding biological sex. Denying that children with atypical sex development exist would be simplistic and wrong. But appropriating the biological reality to make simplistic claims about the nature of sex, such as that sex development is a spectrum, is also a wrong simplification. Sex development follows well-trodden pathways, resulting in clusters of children with similar patterns of sex anatomy and physiology, whether typically male or female, or whether following the biologically explicable, objectively observable, patterns of atypical sex development.

Diversity of sex development means that there are a variety of patterns of sex development; some people are born with typical male anatomy and physiology, some are born with typical female anatomy and physiology, and some people are born with elements of both male and female anatomy and physiology. It may be convenient for some theorists to focus on the intersex experience in order to establish that biological sex is contentious and then extrapolate that contentiousness onto the whole of the population. We contend that respect for people with atypical sex development means acknowledging their particular circumstances and concomitant adversity and not using them as a means to other's ends. Society should accept trans children's bodies without conflating them with atypical sex development. At the same time we should ensure trans issues do not subsume the intersex identity. Acknowledging and respecting the bodies children have is the first step towards heeding the entitlements of all children.

Acknowledgement

I would like to thank Susan Matthews who provided input to an earlier draft of this chapter.

References

Accord Alliance. (2019). *What Is Useful About the Terminology of DSD? What Is Unhelpful?* Retrieved from
http://www.accordalliance.org/learn-about-dsd/faqs/
Blackless, M., Charuvastra, A., Derryck, A., Fausto-Sterling, A., Lauzanne, K., & Lee, E. (2000). How Sexually Dimorphic Are We?

Review and Synthesis. *American Journal of Human Biology, 12*(2), 151-166.
Darlington Statement (2017, March 10). *Joint consensus statement from the intersex community retreat in Darlington.* Retrieved from http://darlington.org.au/statement/
Delisle, R. (2018, March 28). Intersex: When a baby isn't quite a boy or girl. *Today's Parent.* Retrieved from https://www.todaysparent.com/
Dreger, A. (2015). *Galileo's Middle Finger: Heretics, Activists, And one Scholar's Search for Justice.* New York: Penguin.
Fausto-Sterling, A. (1993, March/April). The Five Sexes. *The Sciences,* 20-25.
—. (2000, Jul/Aug). The Five Sexes, Revisited. *The Sciences,* 18-23.
—. (2018, October 25). Why sex is not binary. *The New York Times.* Retrieved from http://www.nytimes.com/
Feder, E., & Karkazis, K. (2008). What's in a name? The controversy of 'Disorders of Sex Development'. *Hastings Center Report 38* (5), 33-36.
Griffiths, D. (2018a). Shifting Syndromes: Sex chromosome variations and intersex classifications. *Social Studies of Science, 48*(1), 125-148.
—. (2018b). Diagnosing sex: intersex surgery and 'sex change' in Britain 1930-1955. *Sexualities, 21*(3), 476-495.
Hausman, B.L. (1995). *Changing Sex: Transsexualism, Technology and the Idea of Gender.* Durham and London: Duke University Press.
Heggi, V. (2015, February 19). Nature and Sex Redefined: We have never been binary. *The Guardian.* Retrieved from https://www.theguardian.com/
Hodgkinson, L. (2015). *From a Girl to a Man: How Laura Became Michael.* London: Quartet.
Intersex Human Rights Australia. (2011). *Basic differences between intersex and trans.* Retrieved from https://ihra.org.au/18194/differences-intersex-trans/
Intersex Society of North America. (2011). *What is intersex?* Retrieved from http://www.isna.org/faq/what_is_intersex.
Kleeman, J. (2016, July 2). 'We don't know if your baby's a boy or a girl': growing up intersex. *The Guardian.* Retrieved from https://www.theguardian.com/
Koyama, E. (2006). From 'Intersex' to 'DSD': Toward a Queer Disability Politics of Gender. *Intersex Initiative.* Retrieved from http://www.intersexinitiative.org/articles/intersextodsd.html.

Merke, D. and Poppas, D. (2014). Management of adolescents with congenital adrenal hyperplasia. *Lancet Diabetes Endocrinolo, 1*(4) 341-352.

Money, J. (1996). Preface. In J. Money & A. Ehrhardt, *Man & Woman, Boy & Girl: Gender Identity from Conception to Maturity*. Northvale, NJ: Jason Aronson Inc.

Money, J. & Ehrhardt, A. (1996). *Man & Woman, Boy & Girl: Gender Identity from Conception to Maturity*, Northvale, NJ: Jason Aronson Inc.

Montañez, A. (2017, August 29). Visualizing Sex as a Spectrum. *Scientific American*. Retrieved from https://blogs.scientificamerican.com/sa-visual/visualizing-sex-as-a-spectrum/

National Academic Advising Association (2014). *Trans Umbrella Resource Sheet*. Retrieved from https://www.nacada.ksu.edu/Portals/0/CandIgender dysphoriaivision/documents/Trans%20Umbrella%20Resource%20Sheet%20ILACADA%202014-1.pdf

National Organization for Rare Disorders (2018). Ovotesticular Disorder of Sex Development. *Rare Diseases*. Retrieved from http://rarediseases.org/rare-diseases/ovotesticular-disorder-of-sex-development

OHCHR (2016, October 26). *End violence and harmful medical practices on intersex children and adults, UN and regional experts urge*. Retrieved from https://www.ohchr.org/EN/NewsEvents/Pages/DisplayNews.aspx?NewsID=20739&%3BLangID=E

Reis, E. (2007). Divergence or disorder?: the politics of naming intersex. *Perspect Biol Med, 50*(4), 535-43.

Sax, L. (2002). How common is intersex? A response to Fausto-Sterling. *J Sex Res, 39*(3), 174-8.

Scottish Trans (2015). Trans Umbrella. *Scottish Trans*. Retrieved from https://www.scottishtrans.org/trans-rights/an-intro-to-trans-terms/transgender-umbrella/

Stoller, R.J. (1974). *Sex and Gender: The Development of Masculinity and Femininity*. London: Karnac.

United Nations Special Rapporteur on Torture. (2013). *Torture and cruel, inhuman or degrading treatments in health-care settings*. Geneva: United Nations.

Witchel, S. (2018). Congenital Adrenal Hyperplasia. *J Pediatr Adolesc Gynecol, 30*(5), 520-534.

CHAPTER SEVEN

BE CAREFUL WHAT YOU WISH FOR: TRANS-IDENTIFICATION AND THE EVASION OF PSYCHOLOGICAL DISTRESS

ROBERT WITHERS

In many ways I would prefer not to have to write this chapter. It could jeopardise my career (see, for example, Brooks et al, 2018). But I believe that many young people have already become sterile, life-long drug dependent medical patients through transgender intervention and many more will yet, if voices like mine which question the wisdom of medical treatment for trans-identified youth, are silenced.

My interest in transgender issues began some twenty-five years ago, when Chris, my first ever transgender client, came to me for analysis (Withers, 2015, 2018a). Chris was a male to female to male de-transitioner. Biologically male, he had fully medically transitioned in his mid-thirties and lived 'successfully' as a woman for nine years before deciding to de-transition. He said he had come to realise he was not really a woman and that surgery and hormones had not solved the problems he hoped they would. He was now trying to live as a man, minus his penis and testes. He still had an artificial 'vagina' that his surgeon had constructed out of a piece of gut, but he had stopped taking oestrogen. His doctors had not yet realised that he needed to take the testosterone his body could no longer produce. Partly as a result of this, he was mired in a deep, apathetic depression. His relationship with his girlfriend was suffering. He was uninterested in sex and unable to orgasm. In addition, he was extremely socially isolated as the previously supportive trans-community now ostracised and vilified him to the point that he felt forced to recant and take down a blog he had posted about his experiences.

Chris desperately regretted that he had not been encouraged to explore his feelings psychologically before undergoing physical treatment. Growing up, his mother had made it clear by her actions that she could only love him if he identified as a girl. His father, who abandoned the family at four, had been a violent and abusive alcoholic. So, Chris had no positive male role model to identify with. He was also autogynephilic: in love with the 'image of himself as a woman', as he put it in the following e-mail sent to me many years after our work together had finished:

> That was me until I orgasmed. Then I loathed myself, the clothes—everything about myself. Shame and self-loathing followed. THAT was when I needed an analyst. I decided to make myself immune. Always after orgasm rage followed. I destroyed whole wardrobes. None of this was ever explained to the psychiatrist at Charing Cross. I had fought hard to suppress the rage and shame, but I was still in a precarious position. However, I passed [as a woman] and that meant I had come to a degree of equilibrium. The hormones and surgery became my goal. I was a natural actor, so it was easy to pass for two years and then become a candidate for surgery. Nothing was ever interpreted. I was never challenged. That is a fearful way for patients to be treated.
> Hope that is helpful. Consider it a gift of gratitude for the man I enjoy being now.

For Chris, trans-identification had appeared to offer a way of evading the shame and humiliation of his autogynephilia. He was given a series of psychological assessments by the gatekeepers at the Charing Cross Hospital. But they were basically 'trans-affirmative' and did not challenge him or attempt to understand the origins of his trans-identification. Meanwhile, a well-meaning 'supportive' trans-community had coached him in what to say at these interviews. As a result, he received the medical treatment he wished for.

In this chapter, I demonstrate how a very similar process is happening for significant numbers of trans-identified young people right now. There is a rapidly growing body of literature (Bonfatto and Crasnow, 2018; Rustin, 2018; Withers, 2015, 2018a; Patterson, 2017; Littman, 2018; Ayad, 2018) suggesting that psychological issues play a crucial role in many young people's trans-identification. Such issues, as these sources show, include difficulties in coming to terms with puberty, other sexual difficulties, social isolation, affect dysregulation, alexithymia, anxiety, depression, self-harm, body dysmorphic disorder, attachment trauma, problems with triangulation and symbolisation, as well as unconscious homophobia and autistic traits. A young person can easily adopt a trans identity as a way of explaining these difficulties, while actually attempting to use that identity to evade them. Online trans activist 'coaches' reinforce the evasion. In England, the past nine years have seen a twenty-five-fold increase in referrals to the GIDS (GIDS, 2018). Once referred to the GIDS, if a young person falls under the care of a trans-affirmative therapist, they can quickly be set on the road to unnecessary, sterilising, lifelong medical treatment.

Meanwhile the scientific research purporting to justify such treatment is often funded by the drug companies who profit from it. In their seminal paper, Delamarre-van de Waal and Cohen-Kettenis (2006), originators of the 'Dutch protocol' on which the current practice of child medical transition is based, acknowledge this link:

> The authors are very grateful to Ferring Pharmaceuticals for the financial support of studies on the treatment of adolescents with gender identity disorders.

The profits can be considerable. The puberty blocker Lupron-PED, for instance, is currently marketed online at $9,233.58 for a three-month supply. Its use remains off-label, but it is often prescribed for several years (Lupron, 2018). The parent company of its current manufactures have already paid an $875,000,000 settlement to the American Department of Justice in 2001 as compensation for corruptly promoting its use in the treatment of prostate cancer, for which it was originally prescribed (Donym, 2018). Biggs (Chapter Two, this volume) presents a cautionary overview of controversies concerning the use of drugs for treating children suffering from gender dysphoria

As I wrote in a newspaper article, in twenty years' time we are likely to look back on this as one of the darkest chapters in modern medical history (Withers, 2018c). Every person with gender dysphoria is an individual, so it is difficult to generalise. From a gender critical perspective, sex is biologically determined while gender is psychosocially constructed. In gender dysphoria, a person's subjective gendered sense of who they are is at odds with the objective biological sex of their body. Many people are able to live creatively with this mismatch and are not in need of any treatment. For those in whom it causes significant suffering however, some sort of treatment may be helpful.

There is no credible scientific evidence of a biological basis for gender dysphoria and the GIDS has stopped screening for one (Tavistock, 2018). In this chapter, I argue it is wise to attempt to work with gender dysphoria psychologically before resorting to irreversible physical treatment. This is especially true of young people, who have so many changes to go through before eventually (if they are lucky) establishing a secure sense of adult identity in their mid to late twenties (Arnett, 2000). Before discussing this further however, I will make some preliminary observations about psychotherapy for gender dysphoric adults and also offer thoughts about detransitioning.

Detransitioning

The quality of research into the treatment of gender dysphoria is generally extremely poor (Donym, 2018; Horvath, 2018; Reynolds, 2018). Nowhere is this clearer than in relation to detransitioning. When James Caspian, a gender psychotherapist, attempted to conduct qualitative research into the subject at the University of Bath Spa, his proposal was turned down on the basis that it might attract negative attention from trans activists and hence bring his university into disrepute (Hardy, 2017). As a result of such fears, the whole area is under-researched, and nobody knows quite how many people regret their medical treatment or why. Prevention of harm and informed decision-making are thus made almost impossible. Unlike most other medical treatments, no double-blind trials, or long-term studies comparing different treatment modalities, have been conducted (Donym, 2018). Instead, potentially damaging, poorly researched medical treatment is offered on the basis that it would be an infringement of a person's human rights to withhold it (Bilek, 2018).

Dr Az Hakeem is a psychiatrist and psychotherapist who ran psychotherapy groups for adults suffering from gender dysphoria from 2001-2012 at the Portman Clinic in London. Speaking of detransitioners and transition regretters in his groups, he says:

> This category of transgender persons was becoming the new silent and marginalised population with no voice or representation. They may feel ashamed about appearing ungrateful to the medical profession for the treatment afforded them- or afraid of letting down or being attacked by the trans community. (Hakeem, 2018a, p.147)

Hovarth (2018) suggests these are some of the reasons why detransitioners and regretters are not involved in follow up studies. I suggest it is possible that those members of the trans community who are most active in silencing and denying the existence of detransitioners are attempting to police in others the doubts they cannot tolerate in themselves. If someone can bear to think about a thing, they can usually bear to let others talk about it, but if a person's sense of identity is built around being trans, talking about doubts and regrets can be experienced as an existential threat (see Withers, 2015 for a case discussion pertaining to this).

Psychotherapy groups for gender dysphoric adults

When Hakeem took over from his predecessor at the Portman Clinic in 2001, there were two very different gender dysphoria groups (Hakeem,

2018a). One group was for presurgical adults who tended to
with hope to their medical transitions. This group often
official medical history designed to gain them access to
treatment they had already decided they needed. The_ .ypically
downplayed psychological co-morbidities and claimed they had always
been trans. It was only by Hakeem stressing a) that he was not a
gatekeeper, b) that he was entirely neutral regarding the wisdom of
medical transition and c) that he was entirely separate from the services
providing such physical treatment, that he was able to elicit a fuller history
of this group's gender dysphoria. The second group had already medically
transitioned but looked back with regret on the transition that had failed to
resolve their dysphoria. 'Unsurprisingly', writes Hakeem 'the resultant
dynamic of this group was that of being stuck in a depressing, helpless and
hopeless position without any prospect of escape' (2018a, p.149).

Hakeem decided to mix the two groups together, which produced remarkable results. About a quarter of his participants had received medical treatment but were still gender dysphoric, while about two thirds were awaiting medical treatment. The rest identified themselves as drag queens, transvestites and so on. Hakeem reports that, after a while, the post-surgical group of regretters became more hopeful. In an Annual Lecture Hakeem also mentioned, almost in passing, that only two of the hundred or so pre-surgical patients he saw went on to medically transition; the other 98% gave up their desire for medical treatment (Hakeem, 2018b). This is extraordinary. According to Hakeem, participants in his groups no longer wished to physically transition once they had encountered the reality of people whose gender dysphoria had not resolved with medical treatment. They still needed the space to talk openly about the doubts stirred up by this experience without jeopardising their own prospects of medical transition. But Hakeem does not appear to have acknowledged the significance of his findings and, as far as I can tell, he has not published them at the time of writing. Perhaps this is because of his determination to remain scrupulously impartial regarding his clients' wishes to medically transition. But to me it seems self-evident that, other things being equal, a life without sterilising and potentially dangerous, expensive hormones and surgery is better than a life with those things.

Children and gender dysphoria

I will now explore in more detail how I believe children are being misled into becoming sterile, drug dependent, life-long medical patients. Where possible, I draw directly on my therapy practice for case material, but the ethics of my profession require me to ask my patients' permission to do so. It could be disruptive for a young person, already struggling with their sense of identity, to encounter my account of their therapy in print. There might be things I would need to say for a reader that my patients or their families would not be ready to hear. For this reason, I will draw on my experience in a general way and, where necessary, create a composite case by linking that experience with pre-existing case material.

I am concerned that children have very often reached the conclusion that they are transgender through a process of self-diagnosis. They may have previously had a feeling that things were not quite right. But some influence, for example a YouTube vlog, TV programme, or someone they know who comes out as transgender, has suddenly helped them make sense of this feeling, and as a result they have self-identified as transgender. After perhaps going on to watch some of the numerous YouTube stars celebrating the benefits of 'top surgery' 'T' and medical transition, they have decided that they too want such treatment. There is no shortage of online 'support' offering them coaching on how to go about getting it (see Lewis, Chapter 12, this volume).

I attended a conference entitled 'The Science of Gender' at the Tavistock Clinic in London (Withers, 2018d). My review of the conference details that it was sponsored by the drug company Pfizer and attended by three representatives of Ferring Pharmaceuticals. There were some very thought-provoking and informative talks, but others seemed to uncritically extol the virtues of pharmaceutical interventions. Any mention of meaningful psychotherapy was conspicuous by its absence. One of the contributors, Professor Kaltiala-Heino, a psychiatrist from Finland, spoke about a phenomenon she terms 'shared identity' (Kaltiala-Heino, 2018). In the gender clinic where she works, she notices numerous young people present identical accounts of their childhood. For instance, several of the biological females she assessed claimed to have spent significant portions of their childhood wandering alone in the forest imagining they were male wolves, which she attributed to online coaching. Some conference delegates posited that this kind of childhood story may assist self-diagnosing adolescents in qualifying for medical treatment. Some fictional histories are clearly just that. But less obvious online coaching can make effective psychological assessment virtually impossible. This can have catastrophic consequences for the patient; as Chris's case shows. A large part of the problem is that self-identification as transgender, which many activists and their allies regard as a human right, has become conflated with the medical diagnosis of gender dysphoria which grants access to medical treatment (Bilek, 2018). Doctors and therapists who counsel caution regarding reliance on self-identification as a tool for diagnosis of transgenderism risk being wrongly perceived as opposed to the human rights of transgender people.

The following case illustrates how an interaction between homophobia and transgender identification invalidates reliance on self-diagnosis and undermines the human rights of LGB people. Alex is a sixteen-year-old, male-identified, biological female who was diagnosed with gender dysphoria by the Tavistock GIDS at the age of 13, at which point she could have been prescribed potentially dangerous puberty blocking hormone treatments (Withers, 2018b, 2018e; see also Brunskell-Evans, Chapter One, and Biggs, Chapter Two, in this volume for an extended discussion of these treatments). Alex says she chose not to take puberty blockers because of concerns about side effects, specifically low mood and fatigue. Alex already suffered from fatigue and a series of other psychosomatic issues such as insomnia, irritable bowel syndrome, lactose intolerance and an eating disorder. I have commented on Alex elsewhere (Withers, 2018b). The following is a description of part of a therapy session:

> He (sic) then spoke of how in the first years of secondary school he tried to fit in and it did not work out. He tried liking the same things as everyone, but it was a 'dark time' in his life; he then found some very good friends who liked the same things he liked and realised he does not have to 'be normal' – and now no-one thinks he is weird because he is a guy. Everyone thought he was really gay, really butch – but because now he is a guy it feels more 'normal'. He spoke of himself as being 'stereotypically a masculine person' and how this was perceived as weird by those around him. Now, I noted, perhaps Alex had got rid of that feeling of 'weirdness' and he has gained some sense of legitimacy as he is now known as trans. Alex nodded in agreement. (Clinical Notes, 2018)

In other words, Alex felt weird as a butch lesbian, but normal once he identified as a straight guy. I would like someone to explain to me how a therapist who simply affirms Alex's trans identification without trying to understand its origins is not practicing gay conversion therapy. And yet some 650 + therapists, counsellors and trans activists wrote in to *Therapy Today* the house magazine of the British Association for Counselling and Psychotherapy, the UK's largest therapy register, demanding:

> Concrete steps to ensure that all of its members are aware that non-affirmative stances towards gender identity are incompatible with membership of BACP (Brooks et al, 2018)

I find it deeply distressing that gay conversion therapy involving the use of sterilising hormones and surgery is effectively being promoted in the name of LGBT rights by such trans-affirmative therapists. Though it is true that other analysts have pathologised homosexuality, to the lasting shame of our profession (Denman 2003), none have advocated sterilising gay people before. While I am not suggesting that anyone is actually doing this, it does seem to have become an unintended—or perhaps unconsciously intended— consequence of trans-affirmative therapy. Alex's self-identification as trans appears to have originated, at least in part, from internalised homophobia. A therapist simply affirming her transgender identification would be colluding with homophobia rather than challenging it, and propelling Alex towards a life as a sterile, straight transman.

From Max to Maxine and back again

It is illuminating to analyse how transgender identification is depicted in popular culture, as it sheds light on the ways that transgender ideology has become normalised, and dangerously misunderstood. In 2018, a three-part TV drama series called *Butterfly* was broadcast in the UK. It depicted the fictional transition of an eleven-year-old boy, Max, who identified as a girl, Maxine, seeking puberty blockers and attempting to cut off his penis. No attempt was made to explain the origins of Max's identification as a girl. My experience and training as a therapist raise serious questions about the likelihood of a child attempting to cut off their penis without deep causal psychological determinants. Three sets of concerns seem to merit careful consideration: (1) What might have happened if the fictional character Max had not received puberty blockers or other medical treatment? (2) What actual path of development does the programme's scenario set Maxine on? (3) How does my therapeutic experience inform a different scenario in which Max is transitioned but comes to understand the psychosocial origins of his transgender identification?

If a child like Max is not given puberty blockers, he is more than likely to spontaneously desist from transgender identification during puberty, and there is also about a 20% chance his dysphoria could persist; the most likely outcome would be he would grow up to be a gay man (Bailey, 2003; Cantor, 2017). As previously discussed in the case of Alex, in my opinion the character Max would probably benefit from non-collusive, exploratory psychotherapeutic support during this process and, like Alex, might find it helped resolve issues underlying his trans-identification. Critics of this approach will probably object that if Maxine does not get puberty blockers to realise an 'authentic self' he is likely to kill himself. The actual suicide figures (Transgender Trend, 2018) show that suicide is extremely rare among self-identified transgender children of Maxine's age. There have only been four suicides in the last ten years in the UK, all of them significantly older than Maxine. These four individuals had not been screened for comorbid conditions such as depression. It seems naïve in the extreme to suggest that feeling suicidal would be better addressed by puberty blockers than psychotherapy or anti-depressants.

To turn to my second question, what is likely to become of Maxine if his future life continued as depicted in *Butterfly*? We know that he has started puberty blockers and is likely to remain on them until the age of sixteen. In the UK the National Health Service may have access to a cheaper generic equivalent, but the cost of Lupron-PED for this period amounts to around $150,000 (Lupron, 2018). Puberty blockers create a 95-100% likelihood of continuation to full medical transition, including surgery (De Vries et al, 2011). The character of Maxine, if he persisted on the medicalised route, would then be unable to have biological children, or experience adult sexual pleasure. We also know (Laidlaw, 2018) that halting Max's puberty would stop the growth of his penis, leaving it too small to turn inside out to create a neo vagina via the surgical procedure usually preferred. Once a neo vagina had been constructed, using a piece of Maxine's gut, it would have to be carefully dilated and disinfected daily because the body would treat it as a wound and attempt to close it up. Nor does the neo-vagina have the mucous membrane or friendly bacteria of a natural vagina (Hakeem, 2018a, p.119).

Turning now to my third question, how does my therapeutic experience inform a different scenario in which Max is transitioned but comes to understand the psychosocial origins of his transgender identification? I argue the problem with the programme *Butterfly* is that it creates unrealistic expectations of medical transition and obfuscates the reality that a person can never actually change sex. The writer of *Butterfly*, Tony Marchant, was advised by representatives of the trans-affirmative UK charity Mermaids. He asks us to believe that Max's identification as Maxine has no cause other than having been 'born in the wrong body'. I draw on my therapeutic experience to suggest a more realistic approach, and also a more genuinely hopeful one, in which the character of Max eventually sees a therapist who helps him to understand the origins of trans-identification instead of just affirming the child's confusions.

In conclusion, this chapter has depicted contrasting approaches regarding trans-identification and gender dysphoria. It is sobering to contemplate that future generations may see current trans-affirmative treatment protocols for trans-identified children and young people, as depicted by *Butterfly*, as a form of 'sexual lobotomy' (Donym, 2018).

References

Asscheman, H. et al. (2011). A long term follow-up study of mortality in transsexuals receiving treatment with cross sex hormones. *European Journal of Endocrinology, 164*(4), 635-642.

Ayad, S. (2018). *How I work with ROgender dysphoria teens* [Online]. Retrieved from http://gdworkinggroup.org/2018/11/12/how-i-work-with-rogd-teens/

Bailey, M. (2003). *The Man Who Would Be Queen: The Science of Gender-Bending and Transsexualism*. Washington: Joseph Henry Press.

Bilek, J. (2018, February 20). Who are the Rich, White Men Institutionalizing Transgender Ideology? *The Federalist*. Retrieved from http://thefederalist.com/

Bonfatto, M. & Crasnow, E. (2018). Gender/ed identities: an overview of our current work as child psychotherapists in the Gender Identity Development Service. *Journal of Child Psychotherapy, 44*(1), 29-46.

Brooks, T. et al. (2018). Letters. *Therapy Today, 29*(3), 17.

Cantor, J. (2017, December 30). How many transgender kids grow up to stay trans? *PsyPost*. Retrieved from https://www.psypost.org/

Delamarre-van de Waal, H. & Cohen-Kettenis, P. (2006). Clinical management of gender identity disorder in adolescents: a protocol on psychological and paediatric endocrinology aspects. *European Journal of Endocrinology, 155*(S1), S131-S137.

Denman, C. (2003). Analytical Psychology and Homosexual Orientation. In R. Withers (Ed.), *Controversies in Analytical Psychology* (pp.157-170). Hove and New York: Routledge.

De Vries, A. et al. (2011). Puberty suppression in adolescents with gender identity disorder: a prospective follow-up study. *Journal of Sexual Medicine, 8*(8), 2276-83.

Donym, S. (2018). *The New Homophobic Bridge to Nowhere: Child Transition*. Retrieved from https://medium.com/@sue.donym1984/the-new-homophobic-bridge-to-nowhere-child-transition-c621d6188d6e

GIDS (2018). *Gender Identity Development Service statistics*. Retrieved from https://tavistockandportman.nhs.uk/documents/408/gids-service-statistics.pdf

Hakeem, A. (Ed.) (2018a). *TRANS: Exploring Gender Identity and Gender Dysphoria*. Newark, UK: Tiger Press.

Hakeem, A. (2018b). *Trans: A Specially Adapted Psychotherapeutic Approach for Gender Dysphoria.* Guild of Psychotherapist's Annual Lecture. Paper presented at the Guild of Psychotherapists, Nelson Square, London.

Hardy, R. (2017, October 17). How a psychotherapist who has backed transgender rights for years was plunged into a Kafkaesque nightmare after asking if young people changing sex might later regret it. *Mail Online.* Retrieved from https://www.dailymail.co.uk/news/article-4979498/James-Caspian-attacked-transgender-children-comments.html

Horvath, H. (2018, December 19). The Theatre of the Body: A detransitioned epidemiologist examines suicidality, affirmation, and transgender identity. *4thWaveNow.* Retrieved from https://t.co/z01IgEb4ti

Kaltiala-Heino, R. (2018, November 18). *Endocrine Perspective- Pubertal development, sexual behaviour, gender identity and mental health in adolescence.* Paper presented at The Science of Gender conference, Tavistock Clinic, London.

Laidlaw, M. (2018, April 5). Gender Dysphoria and Children: An Endocrinologist's Evaluation of I am Jazz. *Public Discourse: The Journal of the Witherspoon Institute.* Retrieved from https://www.thepublicdiscourse.com/2018/04/21220/

Littman, L. (2018). Rapid-onset gender dysphoria in adolescents and young adults: A study of parental reports. *PLOS ONE, 13*(8). doi:10.1371/journal.pone.0202330

Lupron (2018). *LUPRON DEPOT-PED.* Retrieved from https://www.lupronped.com/

Memorandum of Understanding (2017). *Memorandum of Understanding on Conversion Therapy in the UK (Version 2).* Retrieved from https://www.psychotherapy.org.uk/wp-content/uploads/2017/10/UKCP-Memorandum-of-Understanding-on-Conversion-Therapy-in-the-UK.pdf

Patterson, T. (2017). Unconscious homophobia and the rise of the transgender movement. *Psychodynamic Practice* [Online], 1–4. doi:10.1080/14753634.2017.1400740

Reynolds, D. (2018, December 29). The Uncharted Territories of Medically Transitioning Children. *Quillette.* Retrieved from https://quillette.com/

Rustin, M. (2018). Clinical Commentary. *Journal of Child Psychotherapy, 44*(1), 132-135.

Tavistock (2018). The Science of Gender: evidence for what influences gender development and gender dysphoria and what are the respective influences of nature and nurture. The 2018 European Society for Paediatric Endocrinology Science Symposium, October 18-19.

Transgender Trend (2018, October 9). Suicide by trans-identified children in England and Wales. *Transgender Trend.* Retrieved from https://www.transgendertrend.com/suicide-by-trans-identified-children-in-england-and-wales/

Withers, R. (2015). The Seventh Penis: Towards effective psychoanalytic work with pre-surgical transsexuals. *Journal of Analytical Psychology, 60,* 390-412.

—. (2018a). The view from the consulting room. In H. Brunskell-Evans and M. Moore (Eds.), *Transgender Children and Young People: Born in Your Own Body* (pp.181-200). Newcastle upon Tyne: Cambridge Scholars Publishing.

—. (2018b). Clinical Commentary. *Journal of Child Psychotherapy, 44*(1), 135-139.

—. (2018c, November 18). In 20 years we'll look back on the rush to change our children's sex as one of the darkest chapters in medicine. *Mail on Sunday.* Retrieved from https://www.dailymail.co.uk/

—. (2018d, November 1). *Conference Review.* Retrieved from http://gdworkinggroup.org/2018/11/01/the-science-of-gender-evidence-for-what-influences-gender-development-and-gender-dysphoria-and-what-are-the-respective-influences-of-nature-and-nurture-18-19-10-18-the-tavistock-and-portman-nhs-fo/

—. (2018e). Clinical Material. *Journal of Child Psychotherapy, 44*(1), 124-128.

Part Two

Cultural Perspectives

Chapter Eight

'Gender Identity': The Rise of Ideology in the Treatment and Education of Children and Young People

Stephanie Davies-Arai

I am the founder of Transgender Trend, a national organisation of parents of children who have self-identified as transgender. We work to raise awareness of the complex issues which may lead to a child identifying as transgender, and we campaign for responses to be based on robust scientific and clinical evidence. We challenge training guidance and resources which deny biological sex and contest teaching children that they may have been born in the wrong body. In this chapter, I demonstrate the dangers of increased alliance between powerful transgender lobby groups and the major institutions involved in children's lives. Evidence from my own work demonstrates the havoc of this alliance on the bodies and lives of children and young people. I argue ideologically-based medical treatment and education of children is not in their best interests, fails to protect and uphold their human rights, and must be resisted.

The political construction of the 'transgender child'

Transgender and LGBT organisations such as Allsorts, Mermaids, Stonewall, the Gender Identity Research and Education Society (GIRES) and Gendered Intelligence in the UK are not child and youth support groups in any non-partisan sense, but activist organisations at the forefront of shaping public policy and government legislation. They are heavily funded by government departments and lottery grants, and collectively they advise and provide training for the police, the Home Office, the Equality and Human Rights Commission, the Crown Prosecution Service, the Prison and Probation Service, along with the Department of Education, and the National Health Service, including its Gender Identity Service (GIDS). GIRES (2017) clarifies that their political goal is the redefinition of human beings so that a woman is no longer an adult human female but a 'psychosocial category'; a 'gender identity' which may be adopted by both males and females. The erasure of biological sex as a meaningful marker in culture and law constitutes an attack on the status, human rights and protections of women and girls, an agenda also revealed by both Stonewall (Stonewall, 2015) and Gendered Intelligence in their calls to delete the sex-based exemptions in the Equality Act 2010, essentially erasing 'sex' itself as a protected characteristic (Gendered Intelligence, 2015).

The denial of biological sex is not a hidden ideological agenda. For example, the Allsorts Schools Toolkit, promoted on the Mermaids website, gives the following advice on the teaching of Sex Education:

> In labelling the genitals, make it clear that *most* rather than all boys have a penis and testicles and *most* rather than all girls have a vulva and vagina. (Allsorts, 2017, my italics)

Teachers are also openly advised to encourage a boy who identifies as a girl to believe that he was literally born female:

> Some trans pupils and students will need support in developing scripts and responses to questions they may be asked about their transition. This may include phrases such as 'I have always been a boy/girl'. (Allsorts, 2017)

The medical profession, schools and children's organisations now have to operate within a context of increasing political pressure to adopt an essentialist theory of gender which is unscientific and unproven. Accusations of 'transphobia' and 'bigotry' emanating from the transgender lobby create a hostile environment in which to maintain a position of neutrality and professional and ethical accountability (Watts, 2018).

The influence of lobby groups is being felt within health and educational settings through NHS professional development resources and school toolkits and materials for the classroom. A critical examination of training delivered by Stonewall, Gendered Intelligence, Mermaids, GIRES and Allsorts reveals an ideological manifesto to change the world in favour of the erasure of biology and its replacement with gender identity. Teachers, for example, are required to learn a whole new lexicon of words and ideological concepts based on the re-ordering of reality according to queer theory, which considers biology to be a social construct merely 'assigned' at birth and 'gender' to be a concrete, material reality, prenatally present in the brain. In the glossary of 'new genders' created by these organisations, the words which remain undefined and conspicuous by their absence are 'boy', 'girl', 'man' and 'woman'.

Children and young people affected by gender dysphoria are now being viewed and treated within this ideological framework, and children in schools are being educated in the principles of queer theory. The illogicality of this ideology is revealed in the inconsistencies and contradictions contained within the material for schools written by its disciples. According to GIRES:

> Our Gender identity is our psychological sense of *fitting into social categories, typically* of 'man' or 'woman'. (GIRES e-learning module 'Gender Variance', my italics)

GIRES claims that it is this feeling about our social position which defines us as boys or girls, and yet also claims that this feeling is determined pre-socially, existing before birth:

> There is considerable scientific evidence that pre-birth brain development predisposes us to identify in a particular way—typically (but not always) —according to the binary model: boy or girl. (GIRES, 2008)

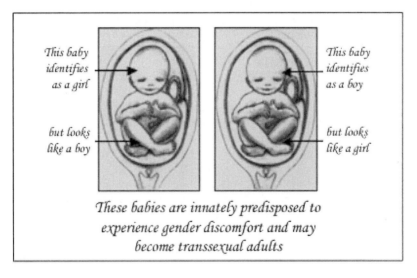

Figure 8-1: Feeling gender in the womb (GIRES, 2008)

GIRES therefore posits that there is something in the developing brain of the foetus which understands social gender rules within a given culture and era to the extent that he, she, or they may be predisposed to reject the binary before birth. Genetic predisposition to certain personality traits is therefore reinterpreted within an 'innate gender' framework before a baby is even born. Inherent in this suggestion is the idea that girls are innately predisposed to accept their position in society and feel comfortable with it and that boys are naturally suited to a position of social dominance.

'Innate gender identity' therefore becomes a means of de-politicising social inequality between men and women and rendering it as the natural order of things. A girl who rejects her position in the hierarchy and experiences great distress in being identified as a member of the lower caste is, by this understanding, really a boy. Using this model, the movement from one sex to the other will inevitably be predominantly one-way, which is what we see in the statistics. When transsexualism was understood through Blanchard's two-type etiology of the autogynephilic and the homosexual transsexual (Lawrence, 2017), adult transsexuals were predominantly male. Since children have been taught that every human being has an innate gender identity and it is this identity which defines the person as male or female, the boy/girl ratio has reversed, and girls now make up over 70% of referrals to the Tavistock clinic.

Gendered Intelligence concurs with GIRES in the innate gender identity model, and yet explicitly describes 'gender' as an outside influence, referring to the 'social shaping' of a boy or a girl:

> Gender: The concepts, roles or attributes that are associated with sex. Gender refers to the social shaping of an individual as being a girl or a boy, man or woman and is represented through behaviour. For example, the behaviour of being aggressive or passive. Gender is often understood as 'cultural'. (Gendered Intelligence, 2008)

Gendered Intelligence goes on to correctly define 'sex' as meaning male or female, but casts biological reality as irrelevant compared to the outside social forces which influence the behaviour of children as boys or as girls. The cultural messages and pressures on children to behave in stereotypical ways according to their sex are elevated by Gendered Intelligence to the position of immutable reality, and social conditioning is recast as an innate 'identity' children are born with. The internalisation of sex-role stereotypes is therefore no longer something to be challenged but to be celebrated as 'your authentic self.' Your 'gender identity' is the true reality.

Queer theory for kids

Transgender lobbyists have worked very hard to translate incomprehensible queer theory into simple, child-friendly language and graphics in order to 'educate' children. Mermaids explicitly embodies the pink/blue brain principle promoted by GIRES in their diagram which represents gender as a spectrum from passive, decorative Barbie to tough, violent GI Joe. Children are encouraged to find where their identity lies along a spectrum between two extremes; according to this model, a boy whose personality aligns more with Barbie than GI Joe must therefore be assumed to have a pink brain and identify as a girl.

One of the most insidious claims of queer theorists is that 'gender' is much more complicated than we had ever imagined and that 'new research' reveals that sex is a spectrum. Although they declare that human sexual dimorphism is a 'social construct', the idea that 'sex is a spectrum' apparently is not. Blinded by science, people presumably feel less confident of challenging the academic posturing of theorists whose new theories conveniently cannot be tested because their claims are based on subjective feelings which are unfalsifiable.

The most ubiquitous teaching resource to educate children in queer theory is the Genderbread Person, promoted by Mermaids and all transgender youth groups in schools, which clarifies the factors a child must now consider in order to find out if they are a boy or girl. The first step is to claim superior understanding of a subject which is so difficult to understand that most people do not:

> Gender is one of those things everyone thinks they understand, but most people don't. Gender isn't binary. It's not either/or. In many cases, it's both/and. A bit of this, a dash of that. (Mermaids, 2018)

Mermaids posits that 'gender' is not 'either/or' but can be 'both', which rather gives away the fact that there is some confusion here about whether or not gender is indeed binary. Next, we get to an arbitrary list of factors a child is being asked to consider in order to be able to understand whether they are a boy, girl, something in between, or neither. This new essential task of childhood and adolescence is referred to by activists as 'exploring your gender identity.' Gender identity is defined as 'your Woman-ness or Man-ness', and is further clarified by this explanation:

> It's how you, in your head, define your gender, based on how much you align or don't align with what you understand to be the options for gender.

It would help if a clear definition of 'gender' had been given first, but in any case children are told that gender identity is 'based on personality traits, jobs, hobbies, likes, dislikes, roles and expectations'. A child has quite a list of factors to analyse before they can be sure whether they are a boy or a girl, and it is not explained which jobs, interests etc mean 'girl' and which mean 'boy' although we can surely guess. Next, we come to 'Gender Expression'—Feminine or Masculine, or:

> The ways you present gender through your actions, dress or demeanor and how those presentations are interpreted based on gender norms.

It is not clear exactly whose interpretation a child must consider, but anyway gender expression is based on 'style, grooming, clothing, mannerisms, affect, appearance, hair, make-up'. Once again it is not explained which aspects of your personal style and grooming indicate that you are a girl or a boy, and the stereotypes children have already imbibed are not challenged. Biological sex is not being 'male' or 'female' but a subjective idea of 'male-ness or female-ness'. There follows a very confused and incomprehensible explanation of biological sex:

> The physical traits you're born with or develop, that we think of as 'sex characteristics' as well as the sex you were assigned at birth [...] Typically based solely on external genitalia present at birth, ignoring internal anatomy, biology and change throughout life.

Finally we learn that sexual orientation is also a piece of the puzzle a child must put together: 'Who You're Sexually Attracted to' and 'Who You're Romantically Attracted to'. This could be: 'Women a/o Feminine a/o Female People' or 'Men a/o Masculine a/o Male People'. Children are told that 'attraction is often categorised within gender' by which they presumably mean heterosexuality, although again this goes unchallenged. The fundamental meaning of sexual orientation is sneakily changed to include attraction towards subjective ideas of femininity or masculinity: the denial that men and women are opposite sexes necessarily entails the denial of sexual orientation. The task, once you have analysed all these factors about yourself, is to put them all together:

> We can think about all these things as existing on continuums, where a lot of people might see themselves as existing somewhere between 0 - 100 on each.

This reduction of men and women to subjective qualities of woman-ness, man-ness, male-ness and female-ness is akin to saying that an object possesses the quality of 'table-ness' when 'table' no longer exists. This material does not just 'confuse' children (which suggests that they may simply be too young to understand it), it fundamentally distorts reality in a way which is harmful for children by conflating 'what you are' (a boy or a girl) with 'who you are' (your personality, your likes and dislikes, your personal style).

Although efforts are being made by lobby groups to convince and persuade children to believe this 'new theory of gender is so complicated most people don't understand it', Jay Stewart, CEO of Gendered Intelligence, suggests that young people came up with it by themselves: 'Young people have a very sophisticated understanding of gender yet the world is lagging behind' (Kleeman, 2015). Polly Carmichael, Director of Tavistock GIDS, has commented: 'Young people are increasingly interested in exploring gender' and 'young people are only just beginning to find and be given their own voice' without questioning *why* children suddenly feel the need to explore or question their gender, nor exactly whose voice it is children are parroting as they all speak from the same script (Gregory, 2015). Bernadette Wren, consultant clinical psychologist at Tavistock GIDS, has observed:

> As more young people come to this service, particularly the teens, they are telling us that it's not the case they feel there are two possible locations for them in terms of gender which is either male or female and the one they were assigned is the wrong one [...] They're telling us there are lots of places on the spectrum in between. That, proportionately, is a small number but it's a growing number of young people. They are saying it's more fluid and it's non-binary and that their destination, at the moment in their teens, is not really settling in the other gender any more than the one they were assigned. (Observer, 2017)

The 'Genderbread Person' has now morphed into the mythical creature of the 'Gender Unicorn' in children's popular culture, which parents report now features in teaching and learning resources in children's schools. This graphic symbolically removes biological sex from the minds of children and young people, introducing the possibility of identifying with the biologically-impossible category of being 'neither male nor female'. The myth that gender identity ideology is some kind of 'youth movement' is a key message of the transgender lobby and a flattering idea to young people themselves, who are characterised as being so much more accepting of diversity and difference. It is clear that this is a youth movement like no other before it: the authentic voice of a generation of young people which originates in the classroom at school and is learned from their teachers.

It is essential to acknowledge that the non-scientific and illogical theory of gender, illustrated by the Genderbread Person and the Gender Unicorn, is the extreme ideology of transgender politics and it is from within this framework of belief that these groups promote the 'affirmation' of a child's gender identity. The acceptance of a child's self-declared identity as a starting point for the safe exploration of feelings and meanings as a standard of therapeutic care is very different to the motivation of activists who promote the affirmative approach from the position that a girl is literally a boy if she identifies as one. There is no question that a child with gender dysphoria must be fully accepted in clinics, communities and schools, and that robust anti-bullying policies and support measures must be in place, however to construct the child as transgender is an ideological stand. Unquestioning affirmation is an approach wholly driven by activists; it is not—as it is presented—a neutral act of kindness and acceptance, it is the active psychological treatment of a child based on an assumption of outcome, and an active adult influence on the child towards that prescribed outcome.

The affirmation approach serves only a political and ideological agenda; it is based on the belief that internally-felt identity trumps biology and it provides the stamp of adult authority and approval on a child's understanding. Lobby groups who campaign for the affirmative approach campaign for the medicalisation of children at ever younger ages: GIRES, Gendered Intelligence and Mermaids have consistently pressured the Tavistock clinic to meet their demands for faster and earlier treatment (Bannerman, 2019). Their agenda could not be clearer: for these lobby groups, affirmation and social transition is the first step on a pathway which leads to puberty blockers and then on to hormones as quickly as possible.

'Life-saving' medical intervention

Trans activists campaign for a system of simple self-declaration of sex in law for adults, without any need for a diagnosis of 'gender dysphoria' or any medical gender reassignment treatment. The question then arises of why they aggressively campaign for earlier and earlier medicalisation of children, lobbying the NHS to reduce the age at which cross-sex hormones may be given. GIRES campaigns for cross-sex hormones to be available 'on an individual basis in accordance with reasonable readiness criteria excluding age', and states that:

We have urged the GIDS to adopt a triage method of assessing the urgency of individual cases and providing fast track access to physical treatment when needed. The triage process could be conducted via telephone and e-mail before the first appointment. Some clients should be seen in less than 18 weeks. (GIRES, 2016)

GIRES' suggestions for the wording of NHS Service Specifications include 'those that enter the service at Tanner stages 3-4 will be fast tracked to start hormone-blockers, bearing in mind that their pubertal changes may be causing them great anxiety'. The idea that natural puberty is too difficult to bear and that denying adolescents access to hormone blockers will cause them to want to take their own lives is consistently publicly promoted by the charity Mermaids: 'medical intervention is very important, especially for teenagers who are already in puberty' says Susie Green, Chair of Mermaids and mother of a trans daughter:

> Medical intervention is very important, especially for teenagers who are already in puberty. It's absolutely vital. If they feel their body is changing against their will, that's when we get a lot of suicidality, self-harm, lots of young people talking about wanting to be dead. If you've got a child who's suicidal and self-harming because their body is changing against their will, nothing is done to fast-track or deal with that need. (Kleeman, 2015)

Bernard Reed, CEO of GIRES, warns of the dangers of not allowing young people to 'be themselves' through preventing their natural puberty and suggests that withholding treatment may cause them to self-harm:

> It can be demeaning and highly stressful for young people not to be themselves. Work suffers as a consequence and they may engage in self-harm. (Marsh, 2015)

Bernadette Wren, consultant clinical psychologist at GIDS, has expressed concerns about the fear being created among parents by these lobby groups, fear which will inevitably be passed on to the children themselves and create the urgency for early intervention:

> It troubles me that parents of very young children are already in terror that their child is going to kill themselves. The energy has to go into changing themselves over at a very early stage because otherwise they'll be tormented to death ... We have shifted to make the treatment available earlier and earlier. But the earlier you do it, the more you run the risk that it's an intervention people would say yes to at a young age, but perhaps would not be so happy with when they move into their later adulthood. (Kleeman, 2015)

The 'trauma' of natural puberty leading to self-harm and suicide has become the established narrative of transgender organisations, a message which is passed on to parents and which for adolescents has become an integral component of the 'true trans' profile for teenagers. Transgender organisations have actively stoked fear to establish 'suicidal ideation' as part of the package of being a transgender teenager. Green describes blockers as 'life-saving':

> The blockers offer the only chance for them to stop the terrible trauma their children have started to go through as they begin to develop into a sex they feel is absolutely alien to them. If you offer hormone blockers at the end of puberty, that is too late. Their body shape will have already changed, and they will have had to live through physical developments that have caused massive distress ... The self-harm and suicide rate among transgender teens is extremely high so offering blockers saves lives. It's quite simple. (Manning and Adams, 2014)

In fact, Green has reported that her own child attempted seven overdoses over a period of three years *after* having had cross-sex hormones at age 12, an earlier age than any other child in the UK (Green, 2018). She is therefore in a unique position to know that there is no evidence to support the contention that early medical intervention will resolve feelings of suicidality in adolescents. There is already evidence emerging of the opposite effect of puberty blockers, outlined in an unpublished GIDS study recently revealed in research by Michael Biggs (2019; see also Chapter Two, this volume):

> What is most disturbing is that after a year on blockers, a significant increase was found ... The study revealed that after a year on GnRHa children reported greater self-harm, and that girls experienced more behavioural and emotional problems and expressed greater dissatisfaction with their body. Tavistock clinicians have begun to speak out on the pressure to 'fast-track' children onto this medical pathway. (Ridler, 2019)

The disappearance of the gender non-conforming child

Another plank of the propaganda disseminated by transgender organisations is that no transgender child desists, that desistance is a myth. When faced with evidence that some children change their minds about transitioning, GIRES suggests they have been misdiagnosed as gender dysphoric by adults:

Children and young people may show behavioural traits that lead adults (parents, clinicians) to draw the conclusion that the child is gender dysphoric, when in fact the child is gender non-conforming for other reasons, including the possible outcome of an LGB identity. These children are not 'desisting', their gender dysphoria does not disappear, they have never experienced it. (GIRES, 2016)

If there is a danger of misdiagnosing the gender non-conforming child, clearly every effort should be made to distinguish them from the truly gender dysphoric. Apart from the fact that blanket 'affirmation' does not allow us to make any distinction at all, gender-affirmative groups in fact do the opposite, consistently conflating 'gender non-conforming' with 'transgender' as if they are the same thing. In their e-learning module for professionals, including teachers, GIRES uses the terms 'trans', 'gender non-conforming' and 'gender variant' interchangeably, suggesting that any child who does not conform to stereotypes for their sex is included as 'transgender' (GIRES, n.d.).

According to Gendered Intelligence, the umbrella term 'transgender' includes 'anyone who may not conform to traditional gender roles' (Government Equalities Office and Gendered Intelligence, 2015). Mermaids, on the About page of their website, even leaves out 'transgender' altogether and makes the claim: 'We work to raise awareness about gender nonconformity in children and young people amongst professionals and the general public', as does GIRES on their What We Do page: 'GIRES' overall aim is to improve substantially the environment in which gender non-conforming people live'. In fact, there is no room for the gender non-conforming boy or girl within the gender identity model. If you are not 'transgender' you are defined as 'cisgender', which means that girls are compelled to define themselves either as people who embrace femininity and accept their social position as girls, or under the trans umbrella as non-binary or some other gender identity. There is no word for the girl who rejects stereotypical notions of femininity; she does not exist. The choice presented here is to be a feminine stereotype or identify out of the 'girl' category altogether; there is no other option.

Children who find it difficult to 'fit in' because they do not meet societal and peer expectations of conformity and do not recognise themselves in the 'cisgender' group are compelled to move to the only alternative group on offer and label themselves accordingly. Is it any wonder that there has been such a steep rise in teenage girls identifying as transgender and that gender clinics are seeing an increase in referrals of highly vulnerable adolescents who may be the most uncomfortable in social gender roles: lesbians, ASD children, those from care homes, those with troubled backgrounds and those who have suffered previous trauma, sexual abuse or mental health issues?

The Politicisation of the Child

We need to be extremely cautious of promoting the idea to children that your identity is the reality and that your body may be an inconvenient mistake. Not only is this idea now being introduced to children from early years and primary education onwards, but it is presented to them as fact. This is a model of understanding which says that your true self is split off from your body and can exist in opposition to it, a model which not only validates and encourages body hatred and rejection but recasts this as heroic and brave. A mind/body split, or disassociation from the body, would, under any other circumstances, be recognised as a symptom of trauma.

Over the past few years we have seen the spread of a global transgender political rights movement which has changed the landscape beyond recognition. Established meanings of words are up for grabs and the battle to re-order society according to the principles of post-modernist queer and identity politics is not only being fought out in the political arena but in the classroom and the consulting room.

Seemingly in the blink of an eye, the reality of the 'child who presents with gender dysphoria' has been replaced with the concept of 'the transgender child'. In this move, both the child and a clinical condition or childhood developmental phase have become politicised. The function of the word 'transgender' changes a child's sex in language and therefore in our fundamental understanding: 'a boy with gender dysphoria' is transformed as if by magic into 'a transgender girl'. For the child, the word used to describe 'myself' changes from embodied reality ('boy') to disembodied identity ('girl') so that conceptually the child now understands that their inner feelings represent reality in the physical world. How does this impact on the child's conception of their own body?

When the child and the body are separated the child becomes a concept to uphold while the child's body is left unprotected. If reality has transferred from the body to the mind, it is the mind's beliefs which are given validation while the body must inevitably be denied. Once the physical body becomes meaningless, devoid of any value, it may therefore be treated with contempt and the child is put at risk.

A child with gender dysphoria may be helped and supported; the 'transgender child' becomes an emblem for a social justice and political rights movement. Children affected by gender dysphoria thus may be used to further a political agenda or provide 'proof' of an ideology. We need to take great care to separate the two issues, both within the medical profession and in schools.

This is not what is happening. It is now clear that within the issue of political rights versus clinical care and safeguarding of children and young people, the waters have become very muddied indeed. The people who are providing professional training for both the NHS and teachers are not clinical or educational professionals, but highly-funded lobby groups with a very specific and extreme political agenda. In the Stonewall schools guidance not once is gender dysphoria mentioned until right at the end in a footnote in the glossary, while the terms 'transgender' and 'gender identity' are used throughout (Stonewall, 2015). The child with gender dysphoria has been disappeared and replaced by the child who embodies and promotes a cause.

The political campaign of transgender activists has two key goals: to establish 'gender identity' as the distinction between men and women in law and to establish 'affirmation' and social transition as the only legitimate approach towards children with gender dysphoria. These two campaign goals are inextricably linked in the furtherance of an ideology which says we are all born with an innate sense of our own gender and it is this internal, subjective feeling which defines us as boys or girls, men or women. The political goal to establish the 'affirmation' approach not only seeks to dictate the treatment of children with gender dysphoria but entails the re-education and indoctrination of all children into the ideology activists campaign to enforce throughout society. The theory of 'innate gender identity' compels every child to re-define themselves within this framework as either 'trans' or 'cis'.

Conclusions

Transgender activist groups influence the areas of medicine, social care, and education for children according to a theoretical model which does not stand up to scrutiny, and yet they have been able to shape policy which affects the health and the lives of all children in negative and potentially harmful ways. I argue the protection of an idea should not take precedence over the protection of human bodies. If human biology is denied and a model of 'innate gender' is adopted, children who identify as transgender are placed outside safeguarding in schools and outside normal duty of care in medicine. In my view, the role of clinicians and teachers is not to validate and reinforce the ideas of one political lobby at the expense of evidence, clinical experience and the safeguarding systems established to protect children: the demands of political activists should not be allowed to put pressure on clinicians, teachers and any other adults entrusted with the care of children and young people.

In my work as Director of Transgender Trend, I witness the devastating impact of gender identity ideology on parents and families. I hear from parents who are told their teenage daughter is now a son, who watch as the whole world tells their daughter she is brave, while they themselves are called transphobic bigots for the following: not wanting her to get her breasts cut off or take hormones which will have serious and irreversible effects on her body; not wanting her to endure a lifetime of health problems as a medical patient with a potentially shortened life-span; or wanting to know the evidence for a treatment so drastic, it may leave her infertile, with uterine and vaginal atrophy leading to hysterectomy, the possibility of excruciating pain during orgasm, and surgical complications if she decides on phalloplasty resulting in possible incontinence, quite apart from the increased risk of various cancers and other serious health problems.

In my work, I hear from parents terrified that their Barbie-and-princess-loving little boy will be led to believe that he was 'born in the wrong body' and that he is really a girl. I hear from parents whose child has already begun questioning if they are a boy or a girl after hearing a transgender presentation in primary school. I hear from parents who say that other parents or even teachers have asked if their short-haired, football-loving little girl is 'transgender' and fear that she will sooner or later pick up on this message that she is not really a girl (Hellen, 2019). I hear from mostly young women who regret their medical transition and the resulting disfigurement of their bodies, and are angry that no adult, including teachers and clinical professionals, ever told them that you cannot actually change sex, that they had been set up to believe a myth.

The 1989 Children Act states that the welfare of the child is paramount. The guiding principle of the UN Convention on the Rights of the Child is that we must act in the best interests of children. The Education Act 1996 protects children from political indoctrination by forbidding the promotion of partisan political views in the teaching of any subject in the school.

I believe guidance and resources for schools produced by transgender organisations and their influence over gender medicine are in breach of these Acts, jeopardise children's human rights and severely compromise professionals in their duty to protect children from harm.

References

Allsorts (2017). *Transgender Inclusion Schools Toolkit*. Retrieved from https://uploads-ssl.webflow.com/5888a640d61795123f8192db/5c0ff2e6f554ac6a3f017600_Trans_Inclusion_Schools_Toolkit_Version_3.3_Jan2019.pdf

Biggs, M. (2019). Tavistock's Experimentation with Puberty Blockers: Scrutinizing the Evidence. *Transgender Trend*. Retrieved from https://www.transgendertrend.com/tavistock-experiment-puberty-blockers/

Bannerman, L. (2019. April 8). It feels like conversion therapy for gay children say clinicians. *The Times*. Retrieved from https://www.thetimes.co.uk/

Gendered Intelligence (2008). *Issues of Bullying Around Trans and Gender Variant Students in Schools, Colleges and Universities*. Retrieved from http://cdn0.genderedintelligence.co.uk/2015/10/07/10-31-21-Trans%20Youth%20Bullying%20Report%20-%20Gendered%20Intelligence%20(formatted%20Aug%2015)%20pdf%20(1).pdf

Gendered Intelligence (2015). *Response to transgender equality inquiry*. Retrieved from http://data.parliament.uk/writtenevidence/committeeevidence.svc/evidencedocument/women-and-equalities-committee/transgender-equality/written/19557.pdf

GIRES (2008). *Transphobic Bullying in Schools*. Retrieved from https://www.gires.org.uk/wp-content/uploads/2017/04/TransphobicBullying-print.pdf

—. (2016). *NHS Service Specification response*. Retrieved from https://www.gires.org.uk/wp-content/uploads/2016/04/GIRES-Young-People-Response-to-Service-Spec-1.pdf

—. (2017). *Report and Accounts for the Year Ending 31st December 2017*. Retrieved from https://www.gires.org.uk/wp-content/uploads/2018/11/GIRES-Accounts-2017.pdf

—. (n.d.). *Caring for Gender Non-Conforming Young People e-learning module*. Retrieved from https://www.gires.org.uk/e-learning/caring-for-gender-nonconforming-young-people/

—. (n.d.). *What We Do*. Retrieved from https://www.gires.org.uk/what-we-do/

Government Equalities Office and Gendered Intelligence (2015). *Providing Services for Transgender Customers: A Guide*. Retrieved from https://www.gov.uk/government/publications/providing-services-for-transgender-customers-a-guide

Green, S. (2018, October 4). Transphobia almost cost my daughter her life. I salute the Guides' inclusivity. *The Guardian*. Retrieved from https://www.theguardian.com/

Gregory, A. (2015, April 7). NHS treating transgender kids aged just three as sex change doctors see soaring numbers of under 10's. *Daily Mirror*. Retrieved from https://www.mirror.co.uk/

Hellen, N. (2019, April 7). Parents in revolt as councils issue schools with transgender toolkits. *The Sunday Times*. Retrieved from https://www.thetimes.co.uk/

Kleeman, J. (2015, September 12). Transgender children: 'This is who he is I have to respect that'. *The Guardian*. Retrieved from https://www.theguardian.com/

Marsh, S. (2015, June 2). From Blake to Jessie: A 7 Year-Old's Transgender Story. *The Guardian*. Retrieved from https://www.theguardian.com/

Lawrence, A. (2017). *Autogynephilia and the Typology of Male to Female Transsexualism: Concepts and Controversies*. Retrieved from http://www.annelawrence.com/autogynephilia_&_MtF_typology.html

Manning, S. and Adams, S. (2014, May 17). NHS to give sex change drugs to nine-year-olds: Clinic accused of 'playing God' with treatment that stops puberty. *Mail on Sunday*. Retrieved from https://www.dailymail.co.uk/

Mermaids (2018). *Genderbread Person*. Retrieved from https://www.mermaidsuk.org.uk/assets/media/Genderbread-Person-3.3-HI-RES.pdf

—. (n.d.). *About*. Retrieved from https://www.mermaidsuk.org.uk/about-mermaids.html

Observer (2017). 'Take these children seriously': NHS clinic in the eye of trans rights storm. *The Observer*. Retrieved from

Ridler, F. (2019, February 17). Doctors at England's only NHS transgender clinic for children warn lobby groups and pushy parents are exposing young patients to 'long-term damage'. *Daily Mail*. Retrieved from https://www.dailymail.co.uk/

Stonewall (2015). *Women and Equalities Select Committee Inquiry on Transgender Equality*. Retrieved from https://www.stonewall.org.uk/women-and-equalities-select-committee-inquiry-transgender-equality

Watts, H. (2018). Helen Watts, Unit leader, Girlguiding UK. *Fair Play for Women*. Retrieved from https://fairplayforwomen.com

CHAPTER NINE

TRANS KIDS: IT'S TIME TO TALK

STELLA O'MALLEY

I work as a psychotherapist in Ireland and write extensively about mental health issues (O'Malley 2017a, 2019). Although I have always been an instinctive feminist, I have no background in feminism or gender politics. In the summer of 2017, I was sitting in a café reading an article about Searyl Atli Doty, an eight-month old baby in Canada whose non-binary parent Kori Doty fought the Canadian authorities for baby Searyl's right to be deemed 'genderless' until the child was 'old enough' to decide upon their gender (Kentish, 2017). As a result of the efforts of the parent, baby Searyl was declared 'U' by the Canadian courts of British Columbia, denoting an 'unassigned' or 'undetermined' gender, thus without male or female designation (ibid). This article lit a fire in me: the infant had been born with apparently perfectly healthy genitals and their physical sex was in no doubt. But Searyl's parent was so entrenched in an ideological position concerning gender that they determined to insist on keeping Searyl's male or female designation off all official records. To me as a parent, Kori Doty's insistence on self-determination of gender for Searyl seemed inappropriate and intrusive. I decided it was time to think more carefully about transgender issues relating to children.

Up until 2017 I was gradually becoming more aware of an explosion in gender politics and trans ideology. Every time I read about children being 'trapped in the wrong body' I would inwardly grimace. I had strong views on this subject drawn from my own profoundly intense experience of being disconnected to my own sex and gender when I was a child. I happened to know from personal experience that a child could move beyond feeling intensely dysphoric one day to feeling completely happy with their gender and their sex the next.

As a direct consequence of reading about Searyl Doty, I wrote an article called 'Boys and Girls: Living in an experiment' for the *Sunday*

Independent about mental health issues pertaining to children with gender dysphoria and wider issues surrounding transgender ideology in general (O'Malley, 2017b). Following this article, I accepted an invitation to be the presenter of a Channel 4 television network team that proposed to explore these issues through a documentary film called *Trans Kids: It's Time to Talk*, broadcast for the first time in the UK on November 21st 2018. In this chapter I will discuss the content of this film and examine the extraordinary obstacles and difficulties we experienced in our attempts to make it. The film aimed to provoke an intelligent, considered and respectful conversation to deepen understanding of the challenges that face children when they identify as transgender, an aim that turned out to provoke intense and sometimes furious debate.

What is gender dysphoria?

According to the UK NHS website, 'Gender dysphoria is a condition where a person experiences discomfort or distress because there's a mismatch between their biological sex and gender identity' (NHS, 2018). Some, but not all people are said to begin to feel this mismatch in childhood. Various indicators of gender dysphoria in children are listed on the NHS website:

Gender dysphoria behaviours in children can include:

- *insisting they're of the opposite sex*
- *disliking or refusing to wear clothes that are typically worn by their sex and wanting to wear clothes typically worn by the opposite sex*
- *disliking or refusing to take part in activities and games that are typically associated with their sex, and wanting to take part in activities and games typically associated with the opposite sex*
- *preferring to play with children of the opposite biological sex*
- *disliking or refusing to pass urine as other members of their biological sex usually do – for example, a boy may want to sit down to pass urine and a girl may want to stand up*
- *insisting or hoping their genitals will change – for example, a boy may say he wants to be rid of his penis, and a girl may want to grow a penis*
- *feeling extreme distress at the physical changes of puberty* (NHS, 2018)

When I was a child, I experienced a profound inner conflict about sex and gender for many years. From as far back as I can remember—from the age of about three years old or so—I wanted to be a boy. I deeply believed I should be a boy. I can easily tick the boxes in the above-mentioned NHS checklist. For me, it did not make sense that I was a girl because I was the strongest kid on my road, I hated everything girly, I was the best fighter, I only played boys' games and I was probably the most boyish kid in the neighbourhood. It was bewildering and very difficult to experience the dissonance of being a girl when I thought I 'should' be a boy. My response was to perform 'being a boy' proudly and assertively, and God help anyone who dared to point out that I was, in fact, a girl.

Everywhere I went I was haunted by the question *'are you a boy or a girl?'* My reaction to this would be to look the questioner in the eye and reply *'a boy'* in a manner so intimidating they would immediately shut up. I was 'persistent, insistent and consistent', which is typically taken as 'the general rule for determining whether a child is transgender' (HRC, 2019). I knew that other girls described themselves as 'tomboys', but I was very disparaging about what I saw as girls making half-baked attempts at being a boy. Deep inside I felt that I was a real boy and tomboys were just girls playing at being boys. Thinking back now, I realise I was a desperately unhappy and lost child who wanted a different life, in a different world, far, far away from the life I was living.

I strongly believe now that, had I been born thirty years later, I would have had everyone around me convinced that I was a 'transgender child'. I would have accessed the internet and obtained hormone medications as soon as feasibly possible. Thankfully, I did not have that opportunity; when I was born in 1974 there was no such construction as a 'transgender child' and I am now a happily married mother of two lovely children. I am also a psychotherapist. Given my personal experience it is little wonder that I became so intensely exercised by the contemporary transgender phenomenon. I wonder whether some of the children who are today described as 'transgender' could be experiencing gender discomfort in a transient way as I did. Can it be that some of the children identifying as transgender believe, as I did, that everything about themselves, their body and their existence is fundamentally wrong and needs to be rejected and maybe even 'fixed'—but will go on to find that their difficulties do not lie in being 'born in the wrong body'?

Beginning the film-making process with trans kids

When initially asked to contribute to a documentary about 'trans kids', I felt my intense childhood experience of gender confusion would add insight to help understand what some children identifying as transgender might be experiencing.

The film-making process started very well. I met with the production team and they all had viewpoints similar to my own. They were liberal, open-minded and wholeheartedly positive about supporting the rights of transgender people. But they were also puzzled about the rising numbers of trans-identified children and had questions about what could be behind the 2500% increase in the last decade in the number of children (and the 4400% rise amongst girls) seeking medical and psychological intervention that could cause permanent and irreversible changes to the body (Tavistock and Portman, 2018; Rayner, 2018). Olly Lambert, the film's director, had already won Emmys and BAFTAs for his work on contentious subjects. When I first met him, he was genuinely uncertain about where he stood on the issue of transgender children but was keen to ask hard, inconvenient questions in a bid to reveal the truth about this issue, no matter how challenging or complicated.

My deepest concern about the film-making process was that we might encounter some lost and lonely children who, as I had once been, were desperately unhappy in their own skin and had become gender dysphoric as a result of their unhappiness. With appropriate therapeutic support, I knew these children could come to self-acceptance through an easier route than transgender identification allows, without medicalising their very existence and without handcuffing themselves to a lifetime of chemical and surgical intervention. I knew this because I had lived through the experience of being disconnected from my gender. I knew first-hand that, for some people, it is possible to come back from believing as a child that you are the wrong sex to growing up to become a person happy with your own gender and at one with your biological sex. It is extraordinary how far people can travel within a lifetime.

We interviewed a range of different people for the film, including Matt, a female to male transgender thirteen-year-old who is also autistic, and twenty-four-year-old Cole, a female to male transitioner who creates vlogs to connect with others who have gender dysphoria. As I delved deeper and deeper into the insider perspectives of interviewees, I realised that the transgender phenomenon raises important societal questions about messages being communicated to children. It troubled me that some children were evidently being sold a false promise of easy, 'off the shelf' fixes to amend their lives and bodies. They were being told they could seamlessly transition to the opposite sex and, one day, 'be' the opposite sex, which would 'make them' the gender of their choice. Of course, the truth is much more complicated; sex transitioning is the beginning of an arduous process of alteration, not the completion of a simple intervention. Gender transitioning has no completion date; there is never a moment of 'arrival' as a person of changed sex because it is not possible for a person to change sex. And gender dysphoria is a difficult and traumatising condition that transitioning sometimes but not always improves, and for which long-term outcomes beyond likely infertility and complex incomplete family dynamics are not known.

From my interviews and research for the film, I realised many transgender children and young people are under the illusion they will one day be a complete version of the opposite sex. This is understandable because 'magical thinking' is a necessary characteristic of children's cognition, enabling the development of creative and divergent thinking (Subbotsky, 2010). However, adults know that life involves a much more fixed set of realities and that complete sex change never happens; a transgender person will be dependent on a life-time of hormone treatments to maintain altered secondary sexual characteristics but will not change sex.

When I met Kenny during the course of making the film, he told me that when he was a young teenage girl nobody ever fully explained the complexities and nuances involved in transitioning to one day become a man. Kenny described how, during transitioning, each appointment would reveal some further information that would add to his knowledge about the transitioning process, and, ten years after the process began, new revelations were still appearing. Initially, when he was a fourteen-year-old girl, Kenny was led to believe that transitioning from a girl to a boy would be a relatively straightforward process and that it would be fully finished at some future point. Only as the months dragged into years did it eventually become clear to Kenny that the truth was a lot more complicated. It was not until turning twenty-one—many years after first taking testosterone and two years after having a full mastectomy—that Kenny realised that 'bottom surgery' to reconstruct the genitals offers more visual than functional outcomes and that he could lose all ability to orgasm if he chose to undergo this surgery. As a consequence to learning this, Kenny then decided not to undergo 'bottom' surgery. And so, Kenny, just like the vast majority of trans men, retains female genitals and reproductive organs (James et al., 2016).

Kenny told me he still suffered period pains, although they are not as dreadfully intense and life-disabling as they were when he lived as a girl. Kenny, nevertheless, said he did not regret transitioning, was philosophical about the incomplete picture of what transitioning involves that he was given as a young teenager, and was remarkably accepting of the slow revelation of uncertain pieces of the transitioning puzzle that he still continues to face. Through using 'Dr Google' one day, Kenny, aged twenty-four, randomly came across the incongruous notion that he could get pregnant.

During the making of the film I also met Cale, a beautiful and thoughtful twenty-five-year-old woman. Cale and I connected on a deep level as we recognised we had come through similarly sad and lonely experiences growing up. Cale and I both had difficult childhoods where we felt alone and disconnected from everyone else. We both felt that we needed to be independent and look after ourselves and to show the world we did not need anyone's love or concern. For Cale and I, a desperate need to prove that we were perfectly fine on our own somehow morphed into an aggressive demonstration of masculinity. We both completely rejected what we perceived as soft, vulnerable femininity and instead strove to be hard, tough and masculine. But Cale is much younger than I am and so she had opportunities I did not have. She had access to the internet and exposure to people who promote transgender ideology and she transitioned when she was a teenager. Cale was on testosterone for many years and had a double mastectomy when she was nineteen.

I learned it is common for transgender people to become obsessed with 'passing' as their preferred gender, and Cale spoke eloquently about how she became obsessed with her ability to 'pass'. 'Passing' in terms of transgenderism is the desire to be perceived as being born in the body of the opposite sex. But this immediately poses the question of 'what's wrong with looking transgender?' When a person becomes intently preoccupied with the idea of passing, this would appear to suggest innate transphobia; if you are horrified at being perceived as transgender, then surely you are transphobic?

Of course, 'passing' involves taking a fundamentally stereotypical view of gender. A transitioning person can only 'pass' if they accept and make use of prevalent sexist stereotypes in their bodily presentation. Perhaps this is why the trans men I met told me that their pecs, their neatly trimmed beard and their strong shoulders demonstrated their manliness, while their vagina and reproductive system were irrelevant. Perhaps this is why breasts are considered a visual signifier of being a woman but details of the reproductive system are less important. Despite talking about internal feelings, I was finding out that in reality transitioning seems to be very focused upon external appearances and upon how the external world judges appearance.

Cale eventually became, as she said in an email to me, 'so very tired of fighting gender roles, social norms and my very own self' and made the decision to detransition. Since detransitioning, her gender dysphoria has completely gone and she said in the film that she now feels much more comfortable in herself and in her own skin. She hopes to one day marry and have children; she seems to have made her peace with the permanent changes that have occurred to her body and hopes the drugs she has taken have not left her infertile.

Shutting down debate

Difficulties making *Trans Kids: It's Time to Talk* started when we moved on from interviewing transgender children and young people to request interviews with representatives of the charities, organisations and advocates who claim to speak on behalf of transgender children and who support their transitions. We wished this to be a balanced film that would show all sides of the debate, and within a few days of talking to young people identifying as transgender I had questions I was burning to ask clinicians and service providers.

I wanted to speak to mental health professionals and advocates for transgender children, to voice concerns building up from what young people were saying and maybe have them allayed. So far in the research process, I had come away with my concerns raised not lowered, and sincerely hoped that there were whole swathes of evidence and information that service providers could reveal that I was not aware of. I cannot stress this enough—like the crew on the film, I was determined to see and hear for myself the 'other side' of this debate, as it was clearly an incredibly complex issue and I was fully ready to be shown information and research that would make me think twice about concerns regarding this issue.

We began by requesting an interview with Dr Helen Webberley, the only private doctor in Britain who provided cross-sex hormones to children (www.mywebdoctor.co.uk). Initially when I spoke on the phone to Webberley, she said she would take part in the documentary but had some stipulations; specifically, she would only be filmed if certain other people were not involved. I silently raised my eyebrows and asked who she had in mind. At first there were two names—the MP for Monmouth, David Davies, and the Director of Transgender Trend, Stephanie Davies-Arai. To clarify, Webberley was not simply refusing to be filmed talking with these individuals. She was refusing to take part in the film if they were to be included in any way whatsoever. This attempt to exclude gender-critical commentators from the film struck me as a profoundly unethical and suspicious request, particularly coming from a doctor who treats children struggling with gender identity. I asked Webberley if there were other people to whom she objected and she said she could not remember off hand and would need to look at the 'little list'. As the film expressly aimed to explore the best way to support and treat gender-confused children, I was genuinely shocked that a doctor was actively trying to silence the views of anyone who had a view that was different from their own. The director of the film said he had never encountered such a request to influence the participation of others in twenty years of filmmaking.

That said, Webberley and I had a cordial conversation. As we had already decided not to interview the two people Webberley mentioned, we uneasily agreed to her conditions and made plans to film with her. But days later Webberley made a sudden about-turn and gave a blanket refusal to let the interview take place. She did not have any new reasons—apparently the article that I had written in the *Sunday Independent* describing my childhood experience was enough to warrant a refusal to engage in this film. Webberley has since been convicted and fined for running an illegal online clinic.

After Webberley's refusal to take part in the film it started to become clear that the film project and I were being shunned by the transgender professional organisations and charities. Susie Green, the CEO of Mermaids (an influential UK charity that claims to support transgender children and their families), directly sought to oppose my work, describing, through a series of emails going back and forth and a lengthy conference call with the production team, that she could not have an interview with me because I had changed my mind about being gender dysphoric—and 'true' transgender kids, according to the insistent Green, do not change their minds.

The question of how we can know who are 'true' transgender children and who are the ones who like me will change their minds was being placed off-limits by transgender lobbyists. I knew it could not be a matter of examining fervency or passion about the subject of transitioning because I clearly remember my own fervent passion. I also remember my own self-loathing and total rejection of my biological sex. Every single peer-reviewed scientific study I located when making the film shows that the majority of trans-identified children eventually desist (Cantor, 2017). Cantor summarised the overwhelming findings of all of these studies in a blog asking 'Do trans- kids stay trans- when they grow up?', and concluded 'no':

> Despite the differences in country, culture, decade, and follow-up length and method, all the studies have come to a remarkably similar conclusion:

Only very few trans- kids still want to transition by the time they are adults. Instead, they generally turn out to be regular gay or lesbian folks. The exact number varies by study, but roughly 60–90% of trans- kids turn out no longer to be transgender by adulthood. (Cantor, 2016)

From the film team's production notes it was clear that Green believed that a documentary which dared to explore the issue of children and young people changing their minds about being transgender would be inherently transphobic. She told the director she would not take part as the film 'was going to be one sided', ignoring the fact that by not taking part she was making such an outcome more, not less, likely. We looked for other contributors to represent transgender support organisations.

All About Trans is a media project that seeks to positively change the way that the media understands and portrays transgender people, and yet their representatives also refused to engage with our documentary. They seemed to be offended that we had not initially checked in with them prior to beginning the filming process. Gendered Intelligence (www.genderedintelligence.co.uk), a prominent UK organisation that provides 'trans youth programmes', educational workshops in schools, colleges and universities and training sessions for teachers and other professionals, did not respond to our calls or emails asking them to take part in the documentary. Like other members of the film's production team I was both surprised and disappointed. I had genuinely expected organisations that exist to educate and inform people about transgender children and their rights would be motivated to contribute to a documentary on the topic, but instead they were closing the door. Something was up.

Next on the list of organisations refusing to take part in the film was Stonewall, the LGTBQI charity, probably the most vocal organisation promoting transgender rights in the UK. The reasons Stonewall representatives gave for refusing to take part were very similar to Green's; they believed it was inherently transphobic to question whether there might be an issue of some children transitioning too early, and made the absurd, false and inflammatory allegation that the film we were making was 'questioning transgender people's right to exist'. And so, refusals to take part in the film multiplied: trans rights campaigners Jane Fae and Jake Graff refused to engage, claiming the presenter (that was me) had 'views' or because they had heard 'troubling things' about the film. Graff would not even speak to the director on the phone. Considering the only 'views' I had expressed were based on my own personal experience of being filled with self-loathing and disconnected from my biological sex when I was a child, I became increasingly enraged by the blanket refusal to engage in conversation about other children and young people's experience of this. What exactly were transgender campaigners afraid of?

Refusals to engage in discussion of any kind about the film went on and on. The journalist and trans rights activist Paris Lees was approached by the Director of Programmes at Channel 4 who personally asked her to take part. Lees arranged to meet the production team but cancelled twice at the last minute and then said she did not have time to participate. Owen Jones, a forthright and opinionated journalist from the *Guardian*, did not respond to emails, and a second approach via a mutual friend came to nothing. Alex Bertie, a usually very media forthcoming trans vlogger, did not respond. Lily Madigan, an assertive male to female transgender campaigner and Women's Spokesperson for the Labour Party in the constituency of Rochester and Strood, would not take part, saying there was 'nothing to discuss'. Member of Parliament Justine Greening, described as 'the driving force behind plans for LGBT-inclusive sex education and gender recognition reform' (Jackman, 2018), declined to contribute to the film.

One of the most frustrating refusals came from Member of Parliament Maria Miller, Chair of the Women and Equalities Select Committee. Miller's committee had published a widely read and influential report on transgender equality, yet she too was persistently unavailable. By now our production team were feeling despondent, angry and suspicious that a co-ordinated campaign to smother the film was underway. We heard of a private WhatsApp group or email chain with leading members from transgender advocacy groups including Webberley and Green putting the word out that no one should take part in the film. We were intrigued about the reasons for this obstruction. The production team had not expressed views on transgender issues. I, as the presenter, was a psychotherapist who had personally experienced intense gender dysphoria as a child. I just wanted to make sure that kids like me were not being overlooked. What exactly was the problem? Why did these powerful transgender organisations and their representatives seek to boycott a film about children and young people identifying as transgender?

Given Miller's influential report and findings, the film's director made a number of attempts to convince her to take part in our film. When told Miller's diary was full, he repeatedly offered alternative dates, ultimately offering to meet Miller anywhere in the UK at any point in the following three months, even for just ten minutes. She finally responded by saying that we should speak instead 'to government' as this was not an issue she was involved with any more. On the very same day however, Stonewall, in partnership with Miller, sent an invitation to MPs inviting them to a panel discussion on transgender equality, confirming that transgender issues were indeed still part of Miller's portfolio.

Polly Carmichael, Director of the Tavistock Clinic (the NHS's largest clinic supporting transgender children in Britain), was the last to refuse to engage in our film about transgender young people. She feared, as noted in a meeting with the production team, that any comment in the film would be 'weaponised' by either side of the debate and therefore do nothing to help the children in her care. I wondered if these kinds of problems are par for the course when making a documentary, but the film's director was even more shocked than I, saying that in almost twenty years of making fifty films, he had never seen or experienced such a co-ordinated campaign to shut down debate.

Clash of beliefs versus science

I realised that there are two sides to a debate on how to support children who experience gender dysphoria. Transgender support groups typically believe that people are born with a 'gender'; for example, that there is some kind of pink soul or blue essence—or even a non-binary, genderless essence—that is somewhere in our bodies. Accordingly, for these organisations, gender is construed as a feeling deep inside our body that has nothing whatsoever to do with social expectations or physical genitalia. By contrast, gender-critical proponents take the view that relying on the physical evidence of our bodies is a reliable indicator of biological sex, while gender is the social conditioning imposed on our biological sex. In these divergent perspectives lies the reason for conflict between 'trans-ideologists' and 'gender-critical thinkers'. In my research for the film I discovered that trans-ideologists feel outraged that gender-critical people do not believe gender is innate; they believe there should be no debate about whether or not gender lives silently within us, and anyone who does not subscribe to this belief is positioned as a 'hater' whose perspective must be shut down. I found gender-critical commentators, in turn, feel outraged that their respect for physical evidence can lead them to be accused of being 'transphobic' or bigoted.

After being obstructed by so many different organisations associated with the transgender community, as described earlier on in this chapter, the film crew then decided to travel to Bristol to watch a debate about women's sex-based rights to have a safe space where panellists would also discuss the ethical issues of attempting to medically sex-transition children and young people. This debate was held in the context of a potential reform of the Gender Recognition Act 2004, which would mean men who identify as women could legally change sex by simply self-identifying as women and thus be enabled to enter all sex-segregated spaces such as prisons, women's toilets and women's refuges from domestic violence. The event descended into chaos. Masked trans rights activists stormed the venue in a bid to set off a smoke bomb and prevent the debate going ahead. Trans rights protestors shouted and screamed the slogan 'trans women are women' for hours in a desperate bid to suppress the debate. In the end, the police intervened to calm the scene and the panel discussion went ahead.

The panel discussion was an intellectual discourse about what it was to be a woman in the twenty-first century and whether certain safe spaces such as prisons and women's refuges were in danger of being cynically infiltrated by opportunists. The discussion also included ethical issues that arise from attempting to medically sex-transition children and young people. Heather Brunskell-Evans, panellist and co-editor of this book, was equated with being a Nazi by the trans activist protesters outside of the meeting for arguing that gender identity is not inherent and that society needs a gentle approach to children's gender confusion. The concerns raised by the panellists that night were later vindicated in media reports, for example, those which reported on convicted rapists such as Karen White who was allowed into female prisons in Britain where 'they' subsequently raped vulnerable women (Gilligan, 2018).

Witnessing these demonstrations and 'no-platforming' demands was so shocking that I did not quite know how to process it. I felt outraged that feminists such as this book's co-editor Brunskell-Evans were being cynically and dishonestly dismissed as transphobic. It was very concerning that trans activists could judge and dismiss genuine concerns without making any attempt to understand the complex points at play. The narrow-minded bigotry displayed by trans activist protesters was deplorable; coupled with the lies, gaslighting and elaborate virtue-signalling, it was enraging. It also gave me some insight into the attitudes of campaigners seeking to 'no-platform' discussants. The young adult protesters I encountered belonged to an organisation called Sisters Uncut. They seemed to me to be genuinely frightened and distressed middle-class students. As I watched them screaming 'trans women are women', my thoughts were that, although it was great to see politically activated young people, their modus operandi depleted their strength as they were arguing that the mere existence of a debate hurt and violated their very existence.

Eventually the film was made and broadcast. After this, one of the trans activists, who was filmed protesting at the event and physically ejecting Brunskell-Evans from the building before the police found an alternative entrance for her, wrote a piece in the *Guardian* acknowledging the protest had achieved nothing and in fact hindered their cause (Betts, 2018). Honest and frank debate, the writer now believed, was the only way forward.

'Nobody has the right not to be offended'

Violently silencing others crosses the line between disagreement and censorship, and with that we move from a free state to an oppressive state. Asking questions, criticising and debating are perfectly satisfactory forms of protest; proposing to plant a smoke bomb is a wholly unsatisfactory form of protest. This is one of the many reasons why pretending that words are 'literally violent' is dangerous—if we treat words as violence then it becomes feasible to sanction a violent response. As I watched attempts at silencing debate through violence during my preparations for a film about transgender children, I could only think it would be so much more empowering for gender-confused children and young people if sensible debate could calmly take place. From a therapeutic perspective, I wondered why transgender activists feel so irrationally and violently scared of people with different viewpoints than their own.

It seems the goal of some transgender activists is total subjugation: they demand silence from everyone who disagrees with them—as, for example, when Webberley demanded the suppression of Davies and Davies-Arai or tweeted to instruct her followers in avoidance of transcritical perspectives 'Do NOT read the newspapers today' (Webberley, 2019), taking up a chilling position as an Executive in Orwell's Ministry of Truth where ideas are beyond question (Brunskell-Evans, 2018). A free exchange of views is treated as 'hate crime'. Any consideration of the therapeutic possibility that, to lead a healthy and functioning adult life, a person unhappy about gender might need support to come to terms with the sex they were born with is responded to as 'hateful'. These tensions became writ large across the film eventually produced.

The fear of inadvertently causing offence

When we finally finished the film-making process, we turned to what would turn out to be a traumatic editing process. Usually, according to the director, the end of filming would have marked the beginning of the end of my role in the film, and he would work with the footage to create a commentary that I would retrospectively read in before the film was finished. But, as it turned out, nothing about this film could be done according to accepted conventions.

A couple of weeks into the editing process I was asked to travel to the UK to record edits, as there were concerns at Channel 4 that the film could be perceived as 'transphobic'. Sore and sensitive after months refuting false allegations about my work, I became immediately enraged; the only issue that mattered was whether the film was transphobic or whether it was not. Why worry about how it might be perceived? I did not know at the time how vulnerable the film would be to vexatious complaints to the UK's communications regulator Ofcom. There followed an exhausting series of re-calls to record compulsory edits, edits and more edits. Every single edit seemed to me to be diluting points I made in the film. Nothing was ever expressly wrong about any of these statements, but it soon became patently obvious that Channel 4 were giving this film a tremendous level of attention. The re-calls and re-edits continued right up until the day before broadcast. Even after previews and advertising of the film, I was still being told that if I did not make the required edits, the film might not be shown. So, I complied with the edits.

Ultra-important urgent edits required the day before broadcast turned out to be yet more qualifying statements: Channel 4 were insisting that when I twice said that there was 'a rise of 2500% in the numbers of children being treated for gender dysphoria', that I had also to say that the actual numbers involved in the 2500% statistic were actually small. They also wished me to point out that the World Health Organisation categorised gender dysphoria as a sexual health issue and not a mental health issue. However, considering there was a lack of consensus on this issue as gender dysphoria was in the DSM 5, I objected to this pointless qualification. These seemed minor points, but Channel 4 were so determined to avert complaints that, only one day before transmission, they were willing to rent out a studio, employ a sound engineer and send me to make these edits. To be sure of transmission, I made the revisions and recorded that the actual numbers that were involved in this 2500% rise were *'from fewer than a hundred to more than two and a half thousand'*, and then at another point in the film provided specific figures, saying *'from 97 to 2519'*. At this stage we were genuinely fearful that the film would not go out and therefore that trans activists would have succeeded in their relentless effort to get this film stopped. In the end, they did not succeed in blocking the film. Despite all the objections, protests, clarifications, input from lawyers, endless emails and multiple edits, eventually the film was aired.

I felt incredibly anxious when the film was broadcast. I dreaded an onslaught of personal attacks on myself and all those involved with the programme. However, the television network's caution returned dividends as nobody—not even trans activists—could feasibly contest any of the points that had finally been made in the film. Everything had been checked, double-checked and triple-checked. Every comment, gesture and look had been edited or significantly amended so that there was nothing in the film that could be described as inflammatory by even the most sensitive reactionary.

Of course, Channel 4's exacting professionalism and sensitivity did not stop malicious and bitter campaigns from people labelling me as 'transphobic' and worse. According to many critics and trolls I would 'never have been labelled a trans kid'. They claimed I did not really have intense feelings, thoughts and beliefs about being the wrong sex, because if I genuinely had, according to transgender doctrine, I would have had no option but to transition. Overwhelming evidence that the vast majority of children grow out of gender dysphoria (Cantor, 2016) is treated as a hateful conspiracy by transgender activists and they will not discuss it.

Another accusation levelled at me was the very serious false allegation that I was trying to debate whether transgender people 'exist'. I do not argue that people who identify as transgender do not exist and I vociferously argue that people who identify as transgender must have the same rights as their non-transgender peers. I do believe we all have the right to ask legitimate questions about people who may regret transitioning and to ask questions about the lack of research on drugs and safety of intervention outcomes. Most of all, we are all entitled and obliged to ask difficult questions to ensure that transgenderism does not harm children and young people.

Conclusion

The entire making of the film *Trans Kids: It's Time to Talk* was embroiled in dishonest, vitriolic and toxic tactics designed to prevent it. I went into the film-making project naively presuming that the desire to reach out to children who felt confused about gender would encourage the transgender activist community to engage. But it did not turn out like that. Instead, I met with people who refused to engage with the issues on any level; I met with people who were afraid to speak out in case they were deemed 'transphobic'; I met with academics who had been silenced; I met with violent protestors who were filled with self-righteous fury; and I met with young people, some identifying as transgender and some detransitioning, who made it plain they do not know who to trust for their support.

I met many wonderful people during the making of *Trans Kids: It's Time to Talk*, but after making the film the world seems a darker and more dangerous place than it did previously. I have discovered that the truth about the experience of transgender children and young people is difficult to explore and extremely hard to present. I learned from Matt, Cole, Kenny and Cale that intervention for children who identify as transgender requires serious examination and debate. I learned from transgender support organisations, activists, politicians and service providers that they will not engage in serious examination and debate.

We must continue to speak out, continue to gather evidence, continue to ask difficult questions and open up every possible way for children and young people identifying as transgender to talk. It's time to talk about transgender kids.

References

Banerji, R. (2012, September 18). Sir Salman Rushdie: Pakistan on the road to tyranny. *BBC World Service*. Retrieved from https://www.bbc.com/
Betts, E. (2018, November 26). I regret my tactics at a trans rights protest. Here's why. *The Guardian*. Retrieved from https://www.theguardian.com/
Brunskell-Evans, H. (2018) The Ministry of Trans Truth. *Spiked*. Retrieved from https://www.spiked-online.com/2018/12/05/the-ministry-of-trans-truth/
Cantor, J. (2016, January 11). Do trans- kids stay trans- when they grow up? *Sexology Today*. Retrieved from http://www.sexologytoday.org/
—. (2017, December 30). How many transgender kids grow up to stay trans? *PsyPost*. Retrieved from https://www.psypost.org/
Gendered Intelligence. (n.d.). *Gendered Intelligence*. Retrieved from http://genderedintelligence.co.uk/
Gilligan, A. (2018, September 9). Why was convicted paedophile allowed to move to a female jail? *The Times*. Retrieved from https://www.thetimes.co.uk/
HRC (2019). *Transgender Children & Youth: Understanding the Basics*. Retrieved from https://www.hrc.org/resources/transgender-children-and-youth-understanding-the-basics

Jackman, J. (2018, January 9). Amber Rudd has been made equalities chief as LGBT reforms remain in limbo. *Pink News*. Retrieved from https://www.pinknews.co.uk/

James, S. E., Herman, J. L., Rankin, S., Keisling, M., Mottet, L., & Anafi, M. (2016). *The Report of the 2015 U.S. Transgender Survey*. Washington, DC: National Center for Transgender Equality.

Kentish, B. (2017, July 3). Parent fights for their child's right to remain genderless. *The Independent*. Retrieved from https://www.independent.co.uk/

National Health Service (2016). *Gender Dysphoria: Symptoms*. Retrieved from https://www.nhs.uk/conditions/gender-dysphoria/symptoms/

O'Malley, S. (2017a). *Bully-Proof Kids: Practical tools to help your child to grow up confident, assertive and strong*. Dublin: Gill Books.

O'Malley, S. (2017b, August 13). Boys and Girls. *Sunday Independent*.

—. (2019). *Fragile: Why we feel more anxious, stressed and overwhelmed than ever, and what we can do about it*. Dublin: Gill Books.

—. (Writer/Presenter), & Lambert, O. (Director). (2018). *Trans Kids: It's Time to Talk* [Television series]. UK: Channel 4.

Rayner, G. (2018, September 16). Minister orders inquiry into 4,000 per cent rise in children wanting to change sex. *Daily Telegraph*. Retrieved from https://www.telegraph.co.uk/

Subbotsky, E. (2010). Magic and the Mind: Mechanisms, Functions, and Development of Magical Thinking and Behavior. *Oxford Scholarship Online*. DOI: 10.1093/acprof:oso/9780195393873.001.0001

Tavistock and Portman (2018). *GIDS referral increase in 2017/18*. Retrieved from https://tavistockandportman.nhs.uk/about-us/news/stories/gids-referrals-increase-201718/

Webberley, H. [MyWebDoctorUK] (2019, April 9). Advice to anyone who cares about the health and safety of our young trans people in the UK. Do NOT read the newspapers today. #transkids #love [Tweet]. Retrieved from https://twitter.com/mywebdoctoruk/status/1115132768571228160

Chapter Ten

Our Voices Our Selves – Amplifying the Voices of Detransitioned Women

Twitter.com/FTMDetransed and Twitter.com/Radfemjourney

Sadly we do not feel comfortable adding our real names to this

Who are we detransitioners?

To understand our perspective as young women who have been through, and pulled back from, identifying as transgender, it's important to know a bit about what we have been through. The two of us who have written this chapter both grew up as non-conforming girls in communities that pushed us into restrictive gender roles when we hit puberty. One of us identified as transgender from the age of 13 to the age of 22 and later re-identified. The other started identifying as trans at the age of 19. She went through a rapid process of transitioning, started on testosterone, had a double mastectomy and then later detransitioned, all by the age of 23. Having changed our minds about transitioning, both of us managed to find a well-hidden community online for people who at some point have identified as transgender. We met in those communities, which have been a tremendous support for us both in making sense of what we went through.

The journey through transitioning to detransitioning and desisting has not been easy. Both of us would rather not have been through it, but we hope our insight into our experience and into the trans community and the detransitioned and re-identified community may shed some light on what goes on in the minds of those going down similar paths as ourselves.

How many are we?

According to a widely cited study by Cecilia Dhejne (2011), only about 0.4% of transitioners will later detransition; we question how accurate this study really is. First of all, Dhejne's study was carried out between 2000 and 2010, which means it was done before the sudden rise of trans-identifying children and young people. This means the study cohort has been superseded by a changed population and nowadays those identifying as transgender are a lot younger than those Dhejne studied, and are therefore likely to have different reasons for their identification. The second problem with Dhejne's research is that in order to be classified as 'detransitioned' in the study, a person would have to get a 'transgender' diagnosis, go through the entire medical process of transition, legally change their gender marker, and then later change their gender marker back. In ordinary day to day life, not everyone who goes under the definition of 'detransitioned' meets these criteria.

A recent survey of 95 subscribes of the online forum r/detrans suggests a significant number of female transitioners identified around 2015 are likely to be detransitioning in the next few years. It will be imperative to plan for adequate health care provision and support (Reddit, 2018). It's hard to put a number on the amount of detransitioners simply because there's a lack of meaningful research on us. What we do know though, through our personal experience, is that there is a social stigma around detransitioning, which makes it very difficult to talk about. Being a detransitioner means feeling like a failure who made a mistake in the eyes of society. The existence of detransitioners poses a threat to the one-sided story of transgender as a permanent reality being promoted by the queer community and trans positive media. Needless to say, to detransition is to make a controversial statement not everyone is willing, able or prepared to make publicly, and so the number of people who detransition is shrouded in secrecy and currently unknown.

Commonalities in the community of girls who once identified as transgender

Through our involvement in the community of girls who have changed their minds about being identified as transgender, we have noticed a pattern of situational similarities which intertwine with each other and underpin our shared experience.

The first common denominator is quite obviously the 'gender non-conforming' girl. She (and we are including ourselves) doesn't fit into the rigid system of gender roles socially ascribed to her, being unable to perform womanhood in what other people tell her is the 'correct way'. Being a lesbian in addition to that hands her an easy way into the queer community. From there on, the pathway to trans-identification is quite short. You don't fit into the expected role of womanhood. Check. You like women. Check. The conclusion is drawn that you're actually a man. Check. This is what happened to us.

Another commonality amongst trans-identified girls is high intelligence. Littman's study (2018) confirming patterns of rapid onset gender dysphoria states that 47% of trans-identifying children discussed in her study were children assessed as gifted. Having a high level of intelligence and at the same time having to fit into restrictive 'rules of girlhood', which typically mean you have to be passive and submissive, does not bode well for a comfortable identity. Growing up, we found that being reflective made us less likely to follow the norm, and gender roles seemed illogical to us, as with anyone with a keen awareness of difference.

We also observe, and personally know, numerous girls identified as transgender who at some point have been subjected to sexual assault and abuse. The detransitioned and re-identified women we know who have been open to us about this tell us that their history of being abused has played a major role in their trans-identification. They say they felt unsafe in their bodies as women, and started to distance themselves from their female bodies as a result of the assault. They say they started to think that if they were born male, they might have been able to avoid assaults and that by becoming male maybe they could avoid future assaults. There is an urgent need for research on links between sexual abuse and transgender identification.

A range of psychological disorders are frequently observed in our community of people who once thought they were transgender, and being on the autistic spectrum stands out significantly (de Vries et al, 2010; Glidden et al, 2016; Littman, 2018). Amongst our community there is no way to avoid the topic of autism spectrum disorders when talking about trans identification; we all recognise social and communication difficulties in our stories. Numerous studies such as those mentioned above show there is a higher percentage of people with ASD diagnoses in the transgender community than in the general population. We think this is likely to be because people with ASD have a harder time figuring out social codes. As the performance of 'femininity' demands coded and socialised behaviours, girls on the ASD spectrum will inevitably experience difficulty with compliance—the high number of autistic girls identifying as transgender suggests they end up thinking they must belong to and identify with the opposite sex instead.

What is common across these factors is that they are all connected to mistreatment of young girls in the patriarchal society we live in. It's neither strange nor a coincidence that the girls who end up identifying as transgender are also generally the ones who have a hard time living up to the expectations of prescribed gender.

The psychological rollercoaster

The traditional image of a trans-identified girl is that she feels terrible because she feels like she is 'in the wrong body' and that the only solution to that despair is medical transition. We know from experience that this line of thinking sets off a dangerous psychological rollercoaster journey and must not go unquestioned. There is an absence of high quality studies evidencing that psychological disturbance is better after transition. There are studies, on the other hand, which show that even though gender dysphoria may be relieved in the initial post-transition period, in the longer term, psychological wellbeing will actually worsen. Dhejne's study, for example, shows the risk of suicide *after* transition is 19 times higher than for the average population.

Based on our own experiences and the stories of others we've come to know through our community, we've made an illustration (Figure 10-1) to map out the track of the desperation felt before transition and the decline in mental health encountered post-transition.

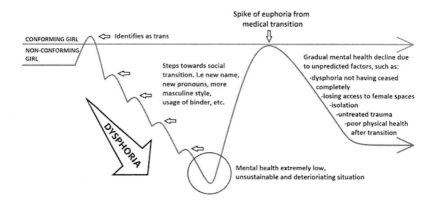

The blue line is a reference line that represents the general wellbeing of a gender-conforming teenage girl. The red line represents a gender-non-conforming teenage girl who identifies as transgender and later starts her medical transition.

Figure 10-1. Our theory on the psychological rollercoaster through transition.

As a gender non-conforming girl you feel as if society does not accept you as you are, you feel uncomfortable navigating your expected gender role and you eventually come to feel there is something inherently 'wrong' with you. Trans-identification initially makes you feel as if you've finally been provided with some answers as to why you've always felt a bit out of place and like you've finally found a space where you fit in. On the psychological rollercoaster, this represents the first positive mental health high in the illustration.

However, as you start getting more comfortable in your new identity, you also start noticing that the environment you are in does not change accordingly. You may feel as if no one really takes you seriously, people may not treat you the way you want them to, they might not use your preferred pronouns and they may be reluctant to view you as the gender you identify as. This means you are now heading into decline, as depicted in Figure 10-1. This leads you to take further steps towards your social transition in pursuit of another climb.

For each step towards transitioning taken, we've noticed that many get a temporary feeling of satisfaction, a small relief, but never a lasting one. In fact, our steps towards transition faced us with constant frustration and disappointment which ultimately intensified criticism of our body, thereby developing a compacted and severe body dysphoria. We both recall thinking 'if only my body was 'right' I would not have to struggle nor be questioned'.

As time passes and you've made sustained efforts to pass as a man without much success, medical transition starts to feel inevitable. Re-identifying as female again does not feel like an alternative because you've been convinced by society and by doctors that gender identity is something you are born with and something that can't be treated without transition. Alternative ways to alleviate dysphoria are not available, nor are they promoted by either the queer community or the health care system.

At this point medical transition feels essential because you're in an extremely desperate and vulnerable position because of the poor state of your mental health and a strong feeling that your situation is not sustainable. Imagining the positive change that medical transition could bring within just a couple of months, with no counter advice whatsoever, in the end makes the decision to start transitioning an easy one. Starting medical transition might relieve some dysphoria, and other people may even start taking you more seriously. You begin to see a possible future without dysphoria, and a chance of finally being seen as the man you feel like on the inside. The one of us who started medical transition definitely felt the mental well-being pitched us upwards again. Unfortunately, in reality, the route to a new identity and psychological well-being wasn't as simple as that.

As we realise now, a lot of trans-identifying girls have other complex issues, diagnoses and trauma on top of thinking they are transgender. But these other issues that underlie gender dysphoria for us—and usually for others—go untreated; wrongfully focusing on transgender identification and beginning medical transitioning does not, of course, mean your other problems cease to exist.

Transitioning into another gender is an undertaking that brings many new problems and difficulties. For women and girls, having to adapt to the male gender role that you have no experience with beforehand is not easy, even though you think you have 'felt like a man' for a long time; suddenly having to blend in and act like one can be confusing and tough. You're faced with 'locker room talk' from an unfamiliar male perspective. You have to re-mould your behaviour into that of 'a man' and part of that also means erasing yourself and parts of your own personality. The deeper you get into transition, the more you're urged to erase your own past, for your own safety as well as for keeping up your own mental image of having always been a man. This means you no longer talk about your time living as a girl and a woman and you no longer talk about the unique experience of being socialised into womanhood. You lose access and connection to female spaces because you now pass as a man. This leaves you feeling isolated, lonely and confused. Following the peak of euphoria reached on the psychological rollercoaster when medical transition first begins, there is plummeting ahead.

We've noticed a new decline in psychological well-being once you start medical transition, because you have physical health issues to contend with that come with medical intervention such as painful scarring, high blood pressure, cardiovascular disease, diabetes and other conditions. Eventually coming to the realisation that you can never actually become a man also leaves you with additional dysphoria that can be hard to cope with. By this time, your rollercoaster ride is not going well.

Some of us who transitioned come to realise at some point that medical transition wasn't the solution to our problems after all. Some who feel they made a terrible mistake in transitioning see no means of escape and are afraid to re-identify or pursue detransitioning. For many there is a real fear of potentially negative repercussions from members of the transgender community who feel detransitioners threaten their identity. In addition, detransitioning brings the dread of seeming like a confused failure to the majority population, so for some this is not an option. A lack of safe ways to get off the transitioning psychological rollercoaster leaves many in a terrible situation with no access to support for detransitioning.

The two of us who have written this chapter are trying to develop support for detransitioning and desisting women by seeking community with others like us, by finding allies, and by trying to raise understanding for others, for example by writing this piece. We see a light at the end of the tunnel now that we've finally found a way to accept our femaleness. We invested so much in believing we could change our situation by trying to change our birth-sex and believing that, in the end, transitioning would make our lives better. Our experiences with identifying and transitioning were not positive and now, having explained why we think this was so, it may be easier to understand why.

From our own experiences and from the patterns we have seen, we do not believe there is any such thing as being 'truly transgender'. We have learned that trans-identification is a result of other issues, both individual and systemic, and that medical transitioning does not improve wellbeing. We believe the affirmative approach we experienced sets children and young people up for a future of distress and confusion, instead of helping them accept and understand themselves better. We think that instead of sanctioning children in their own gender suffering through affirmation, there is a need to figure out the underlying causes for each child's transgender identification to help them deal better with their own situation. By not pushing children towards transitioning, we might begin to tackle the bigger issues underlying gender dysphoria such as gender stereotypes, misogyny, abuse, homophobia, and mental illness.

References

De Vries, A. L., Noens, I. L., Cohen-Kettenis, P. T., van Berckelaer-Onnes, I. A., & Doreleijers, T. A. (2010). Autism spectrum disorders in gender dysphoric children and adolescents. *Journal of Autism and Developmental Disorders, 40*(8), 930-936.

Dhejne, C., Lichtenstein, P., Boman, M., Johansson, A.L.V., Långström, N., & Landén M. (2011). Long-Term Follow-Up of Transsexual Persons Undergoing Sex Reassignment Surgery: Cohort Study in Sweden. *PLOS ONE, 6*(2), e16885. https://doi.org/10.1371/journal.pone.0016885

Glidden, D., Bouman, W.P., Jones, B.A., & Arcelus, J. (2016). Gender Dysphoria and Autism Spectrum Disorder: A Systematic Review of the Literature. *Sexual Medicine Reviews, 4*(1), 3-14.

Littman, L. (2018). Rapid-onset gender dysphoria in adolescents and young adults: A study of parental reports. *PLOS ONE, 13*(8), e0202330. https://doi.org/10.1371/journal.pone.0202330

Reddit (2018). Stats on detransitioners. *Reddit.* Retrieved from https://www.reddit.com/r/detrans/comments/9z4zpa/stats_on_detransitioners/

Resources

twitter.com/radfemjourney runs the blog detrans-identified.tumblr.com.

This resource space shares stories about and by people who earlier identified themselves under the trans umbrella. Submissions from re-identified and detransitioned people are welcomed.

https://www.reddit.com/r/detrans/

This resource space welcomes detransitioners, self-questioners, and detrans allies to post anything about gender detransition: ask questions, share memes, inspire, vent, wonder, etc.

CHAPTER ELEVEN

DETRANSITION WAS A BEAUTIFUL PROCESS

PATRICK

> I seldom have the impression during the first session that it is the real thing. You are in pain!

These were the words that my therapist used some years ago to validate me and confirm a diagnosis of transsexuality. After 45 minutes, I received a letter of referral to start medical transition. At the time, I felt elated, thinking 'I am not delirious and my intuition hasn't failed me! This specialist, a transmasculine person, is bound to know what he is doing. I am transgender!' Did I know what I was getting into in those 45 minutes? No. Two and a half years later, after a nightmarish transition, I understood that this immediate confirmation was only part of a broader evolution in transgender care, now called 'gender-affirming therapy'. My doctor's validation had actually meant nothing. My own psychotherapeutic process during the awful experience of transition helped me understand that the best decision was to drop hormone substitution therapy and detransition, both medically and socially.

Transitioning made me suicidal. There have been several suicide attempts in the process of trying to transition. In order to explain this disastrous outcome, some might argue that I was not transgender in the first place. Nothing could be more complicated. Back then, I had genuinely developed the need to be seen as a member of the other sex. Misgenderings were extremely painful and triggered serious crisis. Moreover, femininity felt natural and suited me perfectly and I longed for a more appropriate body.

Once I started transitioning, I grew suicidal because I started to develop a sense of self which bore no connection with my physical reality. My body wasn't going to become female. I was only going to switch gender roles with medical assistance to help me be more convincing in front of the mirror and in society. I am always amazed by the lack of critical distance with which so many people use the expressions 'male to female' or 'female to male' when in fact there is no such thing as a biological transition. One cannot make a biological woman out of a biological man, just a 'social' version of a woman. The medical path to supposed transitioning is a mere bricolage. Having embarked on medical transitioning it becomes absolutely distressing to notice that one isn't becoming the other sex, but some kind of patchwork, with scars and implants, in a life-long medicalization process with synthetic hormones. Even though the superficial adjustment was quite good in the end, I felt that I was only becoming freakish. I projected a great deal of my despair on the fact that there is no real sex change on passing, which in the trans jargon means the capacity to be read as a member of the other sex without the trans condition being noticeable. Since I was not going to become the other sex, I at least had to look like it.

In what kind of condition was I condemned to live? Undergoing medical transition rendered me stuck in some kind of dead-end identity, with a medicalized body that might look like a woman's body but without me being an actual woman. Obviously, I could appear as a 'social woman'. But deep down I couldn't bear the thought that I was partaking in a definition of womanhood that reinforces sexism: I know being a woman is not a social role! I found my trans condition degrading both physically and socially.

According to the most standard typical definition of trans identity, a trans individual doesn't identify with his or her biological sex. I would put it differently: a trans individual doesn't index his or her identity on biology but on a social gender role and might, in the process, feel the urge to adapt his or her body. I wasn't becoming a woman by virtue of biology, but only as a result of a counter-identification dynamic. A psychological constraint, so to say. One often reads that a trans individual has the intimate conviction of belonging to the other sex. But this modality of existence eventually felt terribly stifling as it was only self-referential: 'I am X because I know that I am X. No questions asked'. I didn't owe my identity to a natural fact of biology shared by other women. I was shutting myself up in a bubble of self-identification. And this turned out to be very dangerous; since my identity had no physical basis, psychologically I regularly collapsed. Suicide attempts ensued. It is awful to live with the obsession of being something without the material basis to support it. I was disconnecting myself from reality and floating between the sexes, with a body that I was only damaging.

I was ruining myself through medical transitioning. Not just physically, but also socially. Not that I experienced much discrimination. But because biology is the criteria of what we call men and women.

Contrary to the transgender doxa that is permeating our culture, sex isn't an assignment. We only happen to name the physical reality which presents itself to our senses and we use language in order to discriminate between different fragments of reality. Of course, some people are intersex, but the existence of this very tiny percentage of the human species doesn't change anything in terms of the overwhelming general rule. Men and women are sexes, not identities assigned at birth. Boys and girls are certainly faced with the cultural pressure to develop, nurture and defend an identity. But this identity doesn't come first. Sex does. I came to realize I need to be what I see when I wake up in the morning, stand up and look at myself in the mirror. Being a woman by virtue of a social performance was not only untying me from a physical determinant, but also placing me at odds with the functioning of humanity. Discrimination as such wasn't responsible for the social exclusion, but the transgender condition itself. I had become dysfunctional through transitioning.

I needed to do a lot of reading to understand the complexity of the trans condition and of the pathways that lead to this syndrome. My first therapist had made no attempt to prevent me from actively developing this condition. Gender-affirming therapy had poured fuel on the fire of my disorder. I understood that I hadn't been treated properly and that I had only been offered a one-fits-all treatment. I also understood the psychological dynamic out of which this cross-sex identification potential had arisen, and also the influence of traumatic events prior to transition. The gender-affirming model of transgender care doesn't allow for a proper examination of the patient's condition and needs. 'Is that your gender? Sure, that can be your gender!' How naïve! How dangerous!

After several suicide attempts and a fortune spent on surgeries, I managed to react, not just on an intellectual level, but also emotionally. I realized I was only suppressing the man that I am *de facto* and trying to become something that I could not be. Sex, I had come to know, is biological and cannot be changed. I might have carried with me a cross-sex identification potential throughout my life, a potential that unfolded under very difficult circumstances and in the context of a personal crisis, but one cannot conflate identity and identification. Medical transition is just a palliative treatment, and a very poor one. Some people might be happy with it, but personally, I ended up finding this solution absolutely monstrous.

******

I had to detransition before it was too late. I do admit that I was scared, as I wondered if this step was going to make me happy or, on the contrary, make things even worse. There had been some suicide attempts and I was still very fragile. What would happen if I felt that I couldn't accept myself as the man that I am? Would this be the end? I have heard from other detransitioners that they have also experienced those feelings. I accepted this uncertainty and proceeded cautiously. I would talk to myself using the masculine form (being French makes it easier). I would drop the voice slowly to see how I would react to a more masculine intonation. I ventured to the men's section of department stores again and switched back very progressively towards a more masculine presentation. One step at a time. I promised myself that detransition shouldn't be a return to the closet, but just an attempt to live in a more adaptive way. I promised myself not to let gender constraints influence me anymore: You are the man that you are. Your body must remain intact, just as nature made you. And your soul is fine as it is. The rest is merely conventional.

Detransition rapidly turned out to be a beautiful process. I was finding again the boy that I was and the man that I grew to be. Old feelings and memories came back up to the surface. I was reconciling myself with the past, a bit like when one comes across old pictures of oneself, filled with a slight melancholy and a sense of indulgence. There had been a great deal of self-loathing.

My body started to recover its natural phenotype. Not immediately though, and my social interactions during the first weeks were quite disconcerting, since the passing had been good enough to be read as a woman or at least as a transwoman. Even with jeans and a t-shirt on, and with no make-up. How ironic! I had achieved so much, but yet I had to detransition. Psychologically, I was rooting myself again in how nature has made me, but I was still perceived as a woman. Fortunately, the reversal of the hormone balance was very fast and I could be read as a man again within approximately two months. This was the proof that I hadn't gone from male to female a tiny bit. Medical transition had only suppressed the natural order of things and created an artificial chemical environment in my body. Nature took over again very quickly.

It was a relief to be able to emancipate myself from a psychological constraint and accept myself as what I am. I even started to appreciate my male body, since it is the only one that I have. The speed of the process was quite exciting. Each week, and sometimes even each day, I would notice the progression of this metamorphosis. The redistribution of the body fat revealed my natural features. The man was re-emerging. I had left the biological plan and lived in a medicalized transgeneric condition for two and a half years. I could root myself again in nature, in a return to some sort of immanence, and overcome this need to transcend a natural given which had locked me in a self-referential modality of existence. Psychologically, it was vital. It was about life. I had to embrace my natural condition.

Using my birth name again was also a beautiful experience. One never pays much attention to the name that our parents have given us. We take it for granted. This name isn't just any name. It was given to me by my parents. Hearing my birth name again after this experience stressed a sense of filiation. I am aware that some readers might find these considerations slightly conservative: the notion of a natural order of things and of filiation can obviously be used in the worst reactionary discourses. In my case, it was just a fact of life that this experience allowed me to appreciate.

Reintegrating the dynamic of men and women was a challenge. I am not the same man anymore. I look at the dynamic between the sexes and within the sexes with a more critical distance. It can be amusing, but also saddening at times. Gender is done to us with a violence that most people hardly suspect. I have overcome some traumas and integrated parts of my soul in what Carl Jung would probably call an 'individuation process'.

I am now more complete as a human being I suppose. However significant, this is probably the only benefit of this process.

Psychotherapy should have been more than enough.

CHAPTER TWELVE

TRANSMISSION OF TRANSITION VIA YOUTUBE

ELIN LEWIS

YouTube transition video blogs known colloquially as 'vlogs' made by 'vloggers', which many parents claim to be the biggest social media platform influencing gender dysphoric young people, are the focus of this chapter. My aim is to understand how much influence YouTube transition vloggers may have over children and young people. Do transition vloggers shape young people's views of their gender identity by providing a particular ideological lens through which to interpret their life and difficulties? Specifically, could YouTube transition videos be playing a causative role in the exponential rise of referrals of gender-confused young women to gender clinics in the UK? In this chapter I look into the world of YouTube transition vlogs, bringing together observations on the language, filters, signs and suggestions embedded in them. I examine their intentions towards, and possible impact on, the minds of young female viewers. To get a sense of this social media platform, I begin by briefly tracing the origins and significance of the YouTube vlogger.

Template of a lonely teenage girl

By June 2006, YouTube had already become established in the internet world of children and young people. Around this time, YouTube videos under the name of Lonelygirl15 began to appear, posted by a 16-year-old called Bree (Lonelygirl, 2006).

Bree was a fun teenager, who described herself as 'a dork', and her videos were pitched in an informal, goofy and intimate style showing her alone in her bedroom, perched on the end of her bed with a floral quilt and a collection of stuffed toys. Hanging on her bedroom door, viewers could see a pink feather boa. Bree talked about everyday teenage concerns such as being lonely and troubles with parents and schoolwork. Her direct conversations to camera were interspersed with speeded up sections of video in which she was seen to clown around and pull funny faces to music. Now and again, she brought her face up close to the webcam and whispered to the viewer. Bree was easy to watch and follow. Viewers were quickly introduced to her boyfriend Daniel. Intrigue was added by hints from Bree that her parents were involved with a strange cult. Within weeks, Bree's YouTube vlogs were a huge hit and teenage fans were hooked.

A few months after launching the vlog, 'Bree' was outed. It turned out the character of Bree was fictitious, her part played by an actress hired by two filmmakers interested in exploring how easy it would be to set up a fake YouTube vlog with a scripted narrative that could evolve over time (Cresci, 2016). Through the use of professional filmic techniques such as quick, frequent cuts and short-burst storytelling, the filmmakers had created YouTube's first viral success of a vlog. The format of LonelyGirl15 became a template for legions of teen bloggers thereafter who copied the format, setting themselves up as one or two characters in a bedroom, chatting informally with their audience about their day to day life, and conspiratorially sharing secrets. Successful blogs still mirror LonelyGirls15's high energy, upbeat friendliness mixed with vulnerability, an invitation to shared intimacy and an absorbing narrative. 'Bree' began each of her videos with a formulaic cheery 'Hi Guys!' or 'Hey Guys', which continues to be one of the most common introductions used by YouTube's 'content creators.' Nowadays, YouTube vloggers have adopted a standardised way of speaking all of their own, dubbed the 'YouTube Voice'. They overstress vowels, add extra vowels between consonants, stretch out vowels and use aspiration to emphasise words—techniques which get and keep the attention of the audience (Beck, 2015). They also speed up and slow down the pace of their speech, use their head and hands a lot, and exaggerate facial expressions such as raised eyebrows and wide open mouths. Liberman, a linguist, likens the YouTube Voice to an 'intellectual used-car-salesman voice', making 'high-energy sales pitches' (Beck, 2015).

The love for YouTube vloggers

YouTube vlogs and vloggers have proved hugely appealing. Surveys have shown that teens 'enjoy an intimate and authentic experience with YouTube celebrities' who are not subject to the traditional, carefully orchestrated images of Hollywood (Ault, 2014). Teenagers are reported as saying they appreciate the candid sense of humour, lack of filter and risk-taking spirit presented by their favourite YouTube stars (Ault, 2014). It is the sense of 'ordinariness' cultivated by YouTube vloggers which makes them 'relatable' (Ault, 2014).

In 2015, a survey of 13-18-year-olds in the US asked children to rate ten popular English language YouTubers and ten popular traditional celebrities in terms of 'influence'. YouTubers took the top eight slots. The celebrity brand strategist who conducted the survey believed the viral nature of promoting YouTube stars across the internet was contributing to their expanding power among teens, commenting that 'YouTube has an inherent ability to create contagious content' (Ault, 2015). There is now strong evidence that vloggers can foster deep and personal connections between themselves and their audience, and that teenagers are developing emotional attachments to YouTube stars that can be as much as seven times greater than those towards a traditional celebrity (Ault, 2015).

Marketeers were quick to catch on to the power of the YouTube vlog and some of the top YouTube vloggers are now reputed to earn millions (Lynch, 2017). Not surprisingly, many young people aspire to follow in their footsteps.

Smartphones and iGen

Online video watching has become part of everyday life for adolescents born after 1995, who are now described as iGen (Twenge, 2017). This cohort has grown up with the internet, social media and smartphones, and iGen's oldest members were only in their early teens when smartphones were introduced in 2007. Smartphones are the most popular device for watching online videos and we know smartphone users are much more likely to consume online videos daily than desktop users (Facebook, 2017). Smartphones have allowed young people to easily consume bite-sized chunks of information—perfect for YouTube videos whose average length is around 4 minutes. Binge watching of video content is a new trend worldwide (Ofcom, 2017), and YouTube's autoplay feature—where a new, related topic video will start playing automatically after a previous one ends—makes binge watching on a smartphone almost inescapable.

iGen teens see less of their friends in real life than previous generations did, and the number of young people who get together with their friends nearly every day dropped by more than 40% between 2000 and 2015 (Twenge, 2017). Today's teenagers are more likely to be found alone in their bedrooms, tuning in to their favourite vloggers, many of whom they consider have a better understanding of them than their real-life friends (Google, 2016). They make virtual friends with others they have never met but about whom they feel they know so much. 'Parasocial relationships psychologically resemble those of face to face interactions but they are of course mediated and one sided' (Chandler and Munday, 2011). Young people who have prolonged exposure to YouTubers can develop intense parasocial relationships with vloggers, and the more they view them, the higher the relational importance of these relationships becomes. Those who both binge watch and subscribe to the regular updates of their favourite vlogger are highly susceptible to being influenced by them. Parasocial relationships are intentionally promoted for the purpose of manufacturing 'markets' by YouTube content creators. The YouTube Creator Academy channel offers tips and techniques on knowing and engaging with an audience, in order to build a 'fan family'. YouTube content producers actively work to elicit parasocial responses in their viewers and measure click rates and comments as key indicators of the intensity of their relationship development with their viewers (Chen, 2014 in Kurtin et al, 2018). Researchers have suggested that embedding health and other behavioural messaging in YouTube content may be extremely effective for educational purposes and propose future studies investigate the impact of messages communicated by YouTube celebrities (Kurtin et al, 2018).

The power of the YouTube vlogger is strong. Is it reasonable, then, to hypothesise that teenage YouTube vlog viewers could be malignly influenced by the vloggers to whom they become so firmly attached? Could YouTube vlogs be used to propagate transgender ideology, for example, to cultivate gender confusion and manufacture 'the transgender child'? At the heart of this new smartphone and iGen world, I infer a problematic potential for undiluted and undue power and influence over children and young people who are confused about gender.

How do transgender vloggers use YouTube?

It is not surprising that transgender vloggers have appropriated the YouTube platform in order to broadcast their experiences and opinions around gender identity formation, gender dysphoria and treatment. Mirroring the rest of YouTube, most successful transition vloggers are white, deemed attractive and in their mid-teens or early twenties. They follow the formula established by the early YouTube vloggers closely: the same 'talking head' style, a format historically used by activist filmmakers to grant expertise to working class women, women of colour and lesbians (Horak, 2014). Their channels typically adopt a story telling a narrative following a hormone timeline, for example, a set of pre-testosterone videos are followed by updates depicting 'One week on Testosterone', 'One month on T', 'Look at me one year on T', sometimes followed by videos celebrating the results of 'top surgery' including 'reveals' in which the vlogger exposes their post-operative body and scars. Such key milestones on the trans journey are interspersed with advice, opinion pieces or general 'checking-in' chats which drive the continuation of the viewer's attention. According to Horak (2014), transition videos are successful because they exploit YouTube's inclination for the 'personal' and the 'spectacular.'

Head and shoulder framing of the camera captures a transman's display of changes to the voice, facial hair, upper-body muscle development and binding or removal of breasts. Cameras zoom in to show physical changes, and time-lapse techniques visually and dramatically compare 'before' and 'after' selves of young vloggers going through transition. 'Highs' and 'lows' of transition are shared, goals are celebrated and a compelling emotional narrative of transformation unfolds before the young viewers' eyes. While there has been much promotion and celebration of YouTube transition vlogs, I am concerned that there has not been a careful assessment of their impact.

Transgenderism goes mainstream

YouTubers had started to vlog about their transition around 2006/7 and by 2010 had cultivated a genre of their own (Raun, 2010). By 2015 an elite group of transgender vloggers had gained hundreds of thousands of subscribers (YouTube, 2015). The rise of the YouTube transition vlog had therefore paved the way for a surge of transition stories to break into mainstream media in late 2014 and early 2015. *Time Magazine* ran an article declaring 2014 to be the 'Transgender Tipping Point' (Steinmetz, 2014). Transgender star Laverne Cox won an Emmy Award and was represented in a waxwork figure at Madame Tussauds, and Caitlyn Jenner's Twitter account launch attracted 1 million followers in just over four hours. Running alongside this, in 2007, the first of the iGen cohort were becoming teenagers and getting smartphones. In parallel, in 2007, annual referrals to the Tavistock GIDS gender clinic began to slowly rise—with a huge, dramatic and unexpected spike beginning in 2015. It seems possible that there could be a connection between the rise in smartphone use, YouTube watching and gender dysphoria rates.

Social contagion, infectious behaviour and gender dysphoria

Social contagion has been defined as 'the spread of affect or behaviour from one crowd participant to another; one person serves as the stimulus for the imitative actions of another' (Lindzey and Aronsson, 1985). Self-harm and suicide rates can vary proportionally according to the extent of intensity and content of exposure to these behaviours, and social contagion is now an accepted risk factor in suicide research (Marsden, 1998; Littman, 2018). As Littman shows, social contagion is also recognised as a factor in anorexia; it seems entirely plausible that it could also be implicated in gender dysphoria. Many parents certainly think so:

> We believe the biggest influence was the online pro-transition blogs and YouTube videos. We believe she was hugely influenced by the 'if you are even questioning your gender you are probably transgender' philosophy. (Littman, 2018)

At a recent conference on child and adolescent mental health, Polly Carmichael, Director of Tavistock GIDS, was asked about a link between social media and the increase in gender clinic referrals. She acknowledged that

> without a doubt there are some young people who are finding a community, friends and all sorts of things through joining a group who have an interest around gender and I think that for some of those we would be very foolish not to acknowledge that it's probably the case that they are caught up in something rather than it being an expression of something that has arisen from within. So there is a lot of concern. (Carmichael, 2018)

Academic and clinical commentators have recently raised concerns that platforms such as YouTube are leading young people to believe that transition is the only solution to their individual gender confusion or discomfort and that the typically uncritically positive depiction of transition may be resulting in adolescents pressuring their families and doctors for medical treatment (Brunskell-Evans and Moore, 2018). Parent-child relationships are reported to deteriorate after a young person starts to identify as transgender, with the young person spending less time with family members and only trusting information about gender dysphoria from transgender sources, usually solely or including YouTube transgender vloggers (Littman, 2018).

Key messages of the YouTube transition vlogs

1. Contested expertise

> Unlike medical professionals, vloggers don't undertake the role of the clinician i.e. by listening to the individual patient's narrative and then making a diagnosis. Instead, they offer mostly generalised advice – taking on the role of educators. There is an asymmetry to this communication, as bloggers rarely respond to viewers' questions. (Dame, 2013)

If a young person asks questions of a YouTube transgender vlogger, they will encounter responses that are general in nature and designed not to alienate the viewer, frequently prefaced with endearments such as 'you do YOU!' or ambiguous palliatives such as 'there's no right or wrong way'. The transvloggers' 'expertise' for advising gender-confused children and young people is assumed by virtue of their own lived experience—their subjective identity is not up for debate. They serve as 'information centres for the questioning', and girls are significantly more likely to have sought out or had contact with trans individuals prior to identifying as transgender (Dame, 2013). Those vloggers who have moved further along the medical transition pathway claim a higher social status within the trans community, creating 'a hierarchy of surgery and testosterone' (Raun, 2016) and an associated hierarchy of vlogger expertise and rank. The more drastic the lived experience of medical intervention undertaken, the more 'qualified' trans vloggers consider themselves to be to advise others. Carmichael (2018) expresses concern about this unfettered expansion of guidance from self-appointed 'experts' whose claimed wisdom is neither assessed nor accredited:

> I have been shocked by some of the things that are swilling around the internet that young people have access to. There are numerous groups on Reddit and Tumblr that many of the young people that are attending our service are going onto ... maybe it's also the dissing of expertise, in a way, so that there is a feeling that this is about who I am, so what does anyone else know? It's a very odd situation in some way.

For self-appointed experts who present YouTube vlogs, there is no requirement for public accreditation of their claims and no accountability. For children and young people who become captivated by the YouTube transgender vloggers, there is only hearsay and induction into the transgender ideology on offer as a resource through which to navigate their personal concerns about gender—and no safeguarding.

2. 'Am I trans?'

The UK Department of Health confirms 'there is no physical test ... for detecting gender variance that may develop into adult dysphoria ... clinicians must rely on the young person's own account' (Department of Health, 2008). There is no requirement to have experienced gender dysphoria since childhood and no age by which one needs to have 'figured things out'. 'In today's social landscape no one really understands the complexity of gender identity development' says the Tavistock GIDS website (GIDS). It is not surprising, given this lack of clinical definition, that 'Am I trans?' is a question frequently asked of trans vloggers. It is also not surprising that 'Am I trans?' is a question transgender YouTube vloggers cannot answer. Some offer a description of their own feelings:

> I always felt incredibly uncomfortable with myself and just like something was wrong, I can't really put my finger on it to explain it better than that, it's just knowing that things don't feel right but not knowing why. (Jammidodger, 2017)

The terms 'gender' and 'sex' are commonly used interchangeably and 'male' and 'female' are used in place of 'masculine' and 'feminine'. The following all-encompassing explanation and instructions are given in a vlog viewed over 267,000 times (at the time of writing):

> In order to be considered transgender ... you have to be someone who does not identify with their birth given sex ... given the fact that gender is a spectrum, you have male on one side, female on the other, and also knowing that transgender just means that you do not identify with your birth given sex, that means you can be transgender and not necessarily feel completely like the opposite gender ... You don't have to be born as a female like I was and necessarily feel like a male all of your life and just know you were supposed to be a boy in order to be transgender. ... So, if you are asking the question am I trans, the answer is probably yes ... how can I figure out where I am on the spectrum? ... Much like science and sexuality the best thing you can do to get results is experiment ... dress like a guy for a month and see how that feels to you, have people start using different pronouns with you and see if you like that ... maybe you are not anywhere on the spectrum, and you're just like your own gender, that's fine, you can still call yourself trans. (Turner, 2015)

Clearly, the vlogger's advice draws young viewers into transitioning behaviour. However, one comment thread below it reveals seven girls in a row (at the time of viewing the comments) who might simply be caught up in 'rapid onset gender dysphoria' and not likely to transition if left to their own devices:

> it kind of bothers me that I have only felt a lot of discomfort with being female for about a year (I'm 17). I have never been a typical girl but being seen as a girl by others never felt wrong to me. It seems so strange to my (sic) that my gender dysphoria has started so late
>
> I am 20 and only recently started feeling like this, but I have heard of people who discovered that they are trans at the age of 40+ and became happy, so I suppose both of us are plenty early!
>
> Same over here (but I'm 15)
>
> Same tho I'm 12
>
> Same. I'm 17 and I am starting to feel my dysphoria since a year ago or so. But it's getting stronger everyday
>
> I have dysphoria ever since I was 12. I'm 13 now
>
> I'm hoping this is normal because I began feeling gender dysphoria this year and last year

A feeling that being 'in the wrong body' stems from a range of comorbidities—such as friendship difficulties, communication problems, trauma and so on—is hidden. For young people seeking an answer to the ordinary discomforts of childhood and adolescence, transgender vloggers offer transition as an allegedly guaranteed pathway to the alleviation of problems. To ensure the induction to transition is secure, the transgender vloggers can provide further instruction on how to obtain a medical diagnosis of gender dysphoria.

3. Telling your story and getting your diagnosis

The diagnosis of gender dysphoria relies on self-affirmation for both the clinic and the young person themselves; therefore, their narrative becomes all important to them. In Littman's parent survey, 69% of parents suspected that when their child first announced a transgender identity, they used language that they had found online: the announcement did not sound like their child's voice and often sounded as if they were reading from a script (Littman, 2018). Young people can learn from YouTube vloggers exactly what they need to say to obtain a transgender diagnosis. Worryingly, this obviates the need for them to come up with their own language and wrestle with conflicting emotions and experiences.

Multitudes of transgender vlogs offer feedback and advice around clinical assessment. Presented in the manner of offering exam tips, the vloggers disclose what questions they were asked and what their replies were. These vlogs act as a guide for young people preparing to visit their own GP or therapist. There are, for example, at the time of writing, over 30 YouTube vlogs from UK girls identifying as male which give accounts of their first appointment with the private clinic of GenderCare in London. These give directions on how to get there, where to find the nearest ATM cash machine for drawing out the clinic fee and a comprehensive run through of the consultation and what they said to achieve the desired outcome of a medical diagnosis of 'being transgender'.

For parents, clinicians and young people, transgender vloggers forge the presentation of young people's confusions about gender and identity. How can anyone be sure of the authenticity of a young person's identification as transgender, especially if—as in most cases—identification as transgender comes out of the blue and following a period of binging on YouTube?

The resource gallery of YouTube transgender vlogs means anyone can learn what to say for a guaranteed diagnosis. Having guaranteed the medical diagnosis, the transition vloggers can then offer advice on how to 'educate' family and friends into concurring with it.

4. How to deal with parents

The education of parents and close family members is a common topic for transgender vloggers. Sometimes parents who affirm their child's conviction of transgenderism make a guest appearance, to be quizzed by the vlogger about how they first felt when learning of their child's transgender identification and how they have supported it since. These appearances are intended to persuade parents who are not endorsing medical transition for their child that affirmation is best. One vlogger's parent, agreeing with the opinion of her transitioned child, says:

> If you have relatives or friends or parents that don't understand it, or are ignorant to it, or can't open their minds to it then basically they don't care, they don't love you ... the ones that don't support you and can't accept it and they think it's weird or anything, cut 'em out. (Sam Collins, 2016)

Certainly, parents who do not ratify their child's identification as transgender are not represented amongst YouTube transitioner vlogs. When a child suggests their own parent is not affirmative, the vloggers position them as intent on their child's unhappiness. Transcritical parents are cast as 'bigoted', 'old fashioned', 'transphobic' and ignorant of the new knowledge of gender identity. Parents are urged to 'listen' to their trans-identifying child and expressly follow the instructions of transgender vloggers. Parents who question their child's transgender presentation, or encourage their child to explore alternative reasons for their bodily discomfort, often find their parent-child relationship deteriorates rapidly (Littman, 2018).

The vlogger ThatGuyOli (2017) invests a GenderCare clinician with somehow knowing more about a young person than their own mother—as does the clinician:

> Parents think that they know everything that is going on with their kids … for the most part you don't know what they are thinking.

Later the mother asks:

> How do you know after 45 minutes with my daughter that this is the right thing for her?

The clinician's reply is recorded as:

> This is an opinion, it's a very subjective thing, there's no MRI scan, there's no test you can do to prove someone is trans, this is just an opinion from a professional.

With a diagnosis and wider social affirmation in place, transition vlogs can next be turned to for the alleviation of profound anxieties about medical intervention.

5. The effects of hormones and surgery

YouTube trans vlogs revere testosterone, colloquially known as 'T', 'Vitamin T' or 'man juice', evacuating it of any harmful connotations. Gender therapist Aydin Olson-Kennedy, Executive Director at the Los Angeles Gender Centre and himself a transman, refers to having had '10 years of testosterone goodness' (MilesChronicles, 2018). Olson-Kennedy invokes the image of a child's plastic toy with removable body parts:

> …there's some folks who want to take testosterone, [or] oestrogen, but don't want to have any surgeries, there's some folks who want to have surgery but no hormones, think Mr Potato Head right, 'I would like a little bit of that but I definitely don't want that'. (MilesChronicles, 2018)

Transmen vloggers focus on external bodily changes that testosterone causes including facial hair, acne, higher temperature, body odour, increased sexual drive, mood changes, muscle development, increased appetite and clitoral growth. The long-term concerns of hormone interventions are downplayed:

> There are speculations that testosterone can lead to an increased risk of certain cancers, but really there is no evidence ... testosterone is not dangerous ... really there are not that many bad things about testosterone, there is the odd change that some people might think is bad, there's the odd health check that you might not want to get, but really overall testosterone is a magic and wonderful thing. (Jammidodger, 2018)

In contrast, advice from Tavistock GIDS clinicians is less encouraging:

> Although puberty suppression, cross-sex hormones and gender reassignment are generally considered safe treatments in the short term, the long-term effects regarding bone health and cardiovascular risks are still unknown. (Cohen-Kettenis & Klink, 2015; Klink et al., 2015 in Butler et al, 2018)

The potential side effects of testosterone use in females are known to include acne, alopecia, reduced HDL cholesterol, increased triglycerides, and a possible increase in systolic blood pressure (Irwig, 2016). There is a lack of high-quality data in the study of testosterone therapy for girls who identify as male, and few prospective or long-term studies (Irwig, 2016). Nonetheless, trans vlogs encourage young viewers to transition without delay.

6. It's all good...

Once identified as transgender, young women and girls are incited to obtain testosterone and get surgery. Vlogs announcing the achievement of these goals carry celebratory titles and a stream of congratulatory comments underneath (Marchiano, 2018). Young people are encouraged to subdue their doubts and push ahead towards what they are told is 'self-actualisation'. Trans vloggers spur them on:

> Stop thinking of your family, stop thinking how anyone else is going to react to you coming out, are you happy with your life right now? Seriously, are you happy with yourself? If not, you need to change something, you need to seriously sit down with yourself and figure it out ... stop waiting around for anyone to give you approval. Being confused and being scared is normal but do not let fear control your life. You're gonna live a sad life

if you let fear control your life, for real. (Taylor, 2018)

> Just do it. Just be the best you can be. Just be whatever the hell that is. As long as you're not hurting anyone else, you're fine. (McKenna in Lorenz, 2018)

> If you want a beard, just go out and get your beard. (TheRealAlexBertie, 2014)

Gender therapist Aydin Olson-Kennedy, who identifies as a 'transgender queer man who happens to be married to a straight cisgender woman', says in a vlog viewed nearly half a million times (at the time of writing) that gender therapy is for anyone having 'something happening with their gender'. When the young vlogger admits to a fear of identifying as something which then might change, Olson-Kennedy provides reassurance:

> We can't actually know how we feel about something until we're there, and so, um, helping people relieve themselves from like 'but I'm so nervous about blah blah blah' and it's like 'we'll totally figure out when we're there'. (MilesChronicles, 2018)

Vloggers often say transition enables them to 'inhabit a new, warmly illuminated world. Images of sunrises and metaphors of being 'born again' are common, as are expressions of gratitude' (Horak, 2014). Transition is 'often articulated as a birth or re-birth signalling a new start in life and new identity' (Raun, 2010). Describing children on the autistic spectrum, Marshall (2018) says:

> Some of them have said to me that they have watched famous YouTubers who have transitioned and their life just looks fabulous ... they are following these people who seem to have wonderful lives on YouTube ... and it reinforces them to proceed with taking hormones and transitioning.

Whilst transitioning may produce an initial euphoria and relief from worries, there is little research on the long-term outcomes (Littman, 2018). Slowly, some young people who have experienced transition are finding the courage to speak a different truth. One vlogger recently described dysphoria as shifting around the body after medical intervention and lamented that this is a problem rarely aired:

> I think dysphoria worsening after surgery is something that a lot of people don't really talk about—I've never really seen a huge discussion about this. I've never really heard people say 'I've got top surgery and now I feel like

shit' and I think that's the thing about the human brain, once you've fixed one problem in your life, another problem arises ... I think we should have a more open discussion about how dysphoria can be worsened by medical transition because medical transition is not this big saving thing. It isn't something that is a cure. Young trans people and people earlier in their transition should know that. It's a harsh reality to face but I think I would rather have faced it sooner rather than later. (JakeFtMagic, 2018)

Another popular female to male transitioner is bravely beginning to vlog about ongoing struggles with depression, not alleviated by hormones, surgery or wide social affirmation:

...realistically I've probably been depressed ... since I was like 16, maybe 15, ever since I started to kind of understand like my gender identity and like probably maybe even my sexuality and the negative repercussions of that like getting bullied and just feeling crap about myself in general ... but I think because I knew I was trans, I never even considered depression a possibility, I always thought it was my gender dysphoria just making me feel terrible and I thought, you know, once I was on testosterone I'd be happy but then suddenly I was on testosterone and I was still unhappy and then I was thinking, OK, once I've had surgery surely then I'll be truly happy, then I had surgery and for a while I was happy. (TheRealAlexBertie, 2018)

These admissions from young people are chilling evidence that medical intervention is not a cure-all for childhood or adolescent distress.

7. Don't delay: Transition or Die

Transition vlogs promote the belief that delaying transition will have detrimental effects on young people's mental health and future life happiness. The vloggers carry a menacing warning about delayed intervention:

Transitioning saved my life ... it allowed me to live as myself and actually live my life and had I had to continue living as female, feeling as uncomfortable and wrong as I had, then, to be honest ... I don't really know where I would be and it's kinda scary to think about. (JammiDodger, 2017)

Without taking my first steps in medical transition I would not be here right now ... that's something very important, specifically like parents of transgender kids need to see. It's very very controversial to put kids on cross-sex hormones, it's very controversial to start hormone blockers, it's very controversial to even be trans as a kid ... so maybe this is a video that you show to your parents. (Rider, 2018)

The vlogger admits to feeling bad about pushing their parents towards an acceptance of medical intervention at such a young age, yet advises others to do the same:

> I pushed my parents harder than I have ever pushed my parents ... because I was so sick of not living my authentic self and I told them straight to their faces that they could not expect me to keep staying around if I could not do the thing that I needed to do. ... It doesn't matter if you're a minor, like if you have to make changes in your life, like if you have to do massive things that you know minors don't usually do or don't usually have the consent of doing, you have to push for it... (Rider, 2018)

Gender-confused teens and their parents have come to have a deep fear that transitioning delays can ultimately lead to suicide, but the data does not back this up (Biggs, 2018). Clinicians from the Tavistock GIDS record that parents and young people are increasingly anxious about the risk of suicide, but nevertheless advocate keeping options open and not immediately rushing to intervention. They state that 'keeping options open is important to allow a young person to feel able to change paths if they want to' (GIDS). Young people's testimonies on the GIDS website bear out the importance of this. Robin says:

> When I started to feel dysphoric of my body, it got very bad, very quickly. It went from a mild to severe hatred of my body and I wanted hormones and surgery to make myself feel less bad about myself. ... It gradually lessened as I went to GIDS and as I was taking antidepressants prescribed by my psychiatrist ... I came to GIDS wanting everything changed about me, and now starting to get to the end of my GIDS journey, not wanting anything changed about me physically. I'm shocked and many others that I'm close to are shocked that I have completely changed my outlook on myself and what I want for the future.

By its nature and intention, the transition vlog deliberately eliminates any exploration of the context of a transgender identity. Vloggers are constrained both by the need to validate their own decisions and by the income generation possibilities of growing their channels via subscribers. They know what their audiences want to see and hear. Surgery and testosterone updates can increase viewing rates tenfold and some videos can garner up to half a million views.

Conclusion

Vloggers have a powerful influence over their audience. As transgender issues entered the mainstream media around 2014, young women watching on their smartphones were reported by their parents to be binge watching FTM transition blogs—and consequently self-identifying as transgender on the basis of non-specific symptoms. By September 2018, the UK Equalities Minister launched an urgent inquiry into the number of teen girls being referred to the UK's only children's gender clinic, the Tavistock Gender Identity Development Service (GIDS). A source at the Government Equalities Office said 'There has been a substantial increase in the number of individuals assigned female at birth being referred to the NHS. There is evidence that this trend is happening in other countries as well' (Rayner, 2018).

I call for an in-depth exploration of the content of YouTube transition vlogs to provide data for all future enquires. Parents, clinicians and researchers are aware of the influence of transitioning vlogs on children and young people and yet, incredibly, to date there has been little public or clinical acknowledgement of this phenomenon. In every other area of a child or young person's life we know the importance of social media on their identity, and we aspire to safeguard against harmful influence. It is unconscionable that in this area of transition vlogs, central to understanding the intervention of transgender children and young people, crucial impact data has not been collected and analysed.

References

Ault, S. (2014, August 5). Survey: YouTube Stars More Popular Than Mainstream Celebs Among U.S. Teens. *Variety*. Retrieved from https://variety.com/

—. (2015, July 23). Digital Star Popularity Grows Versus Mainstream Celebrities. *Variety*. Retrieved from https://variety.com/

Beck, J. (2015, December 7). The Linguistics of 'YouTube Voice'. *The Atlantic*. Retrieved from https://www.theatlantic.com/

Biggs, M. (2018). Suicide by trans-identified children in England and Wales. *Transgender Trend*. Retrieved from https://www.transgendertrend.com/suicide-by-trans-identified-children-in-england-and-wales/

Brunskell-Evans, H. and Moore, M. (Eds.) (2018). *Transgender Children and Young People: Born in Your Own Body*. Newcastle upon Tyne: Cambridge Scholars Publishing.

Butler, G., De Graaf, N., Wren, B., and Carmichael, P. (2018). Assessment and support of children and adolescents with gender dysphoria. *Archives of Disease in Childhood, 103*, 631-636.

Carmichael, P. (2018). *The Association for Children and Adolescent Mental Health Emanuel Miller Lecture.* Paper presented at the Emanuel Miller Memorial Lecture & National Conference, March 16. Available at https://twitter.com/acamh/status/974676428972220417

Chandler, D. and Munday, R. (2011). *A Dictionary of Media and Communication.* Oxford: Oxford University Press.

Chen, C-P. (2014). Forming digital self and parasocial relationships on YouTube. *Journal of Consumer Culture*, 1-23. doi: https://doi.org/10.1177%2F1469540514521081

Cresci, E. (2016, June 16). Lonelygirl15: how one mysterious blogger changed the internet. *The Guardian.* Retrieved from https://www.theguardian.com/

Dame, A. (2013). "I'm your hero? Like me?" The role of 'expert' in the trans male vlog. *Journal of Language and Sexuality, 2*(1), 40-69.

Department of Health (2008). *Medical care for gender variant children and young people: answering families' questions.* Retrieved from https://www.mermaidsuk.org.uk/assets/media/GIRES%20doh-children-and-adolescents.pdf

Facebook (2017). *Moving pictures; the persuasive power of video.* Retrieved from https://www.facebook.com/business/news/insights/moving-pictures-the-persuasive-power-of-video

GIDS (n.d.). *Gender Identity Development Service.* Retrieved from http://gids.nhs.uk/

Google (2016). *Why YouTube stars are more influential than traditional celebrities.* Retrieved from https://www.thinkwithgoogle.com/consumer-insights/youtube-stars-influence

Horak, L. (2014). Trans on YouTube: Intimacy, Visibility, Temporality. *TSQ, 1*(4), 572-585.

Irwig, M.S. (2016). Testosterone therapy for transgender men. *The Lancet.* DOI: https://doi.org/10.1016/S2213-8587(16)00036-X

Kurtin, K.S., O'Brien, N., Roy, D., and Dam, L. (2018). The Development of Parasocial Relationships on YouTube. *The Journal of Social Media in Society, 7*(1), 233-252.

Littman, L. (2018). Rapid-onset gender dysphoria in adolescents and young adults: A study of parental reports. *PLOS ONE, 13*(8), e0202330

Lindzey, G. and Aronson, E. (1985). *Handbook of Social Psychology: Group Psychology and the Phenomena of Interaction* (3rd Ed.). New York: Lawrence Erlbaum Ass.
Lorenz, T. (2018, June 22). Transitioning on YouTube: While the platform's other stars pull pranks in the hopes of going viral, Miles McKenna is helping fans figure out who they are. *The Atlantic*. Retrieved from https://www.theatlantic.com/
Lynch, J. (2017, December 8). Meet The Youtube Millionaires: These are the 10 highest-paid YouTube stars of 2017. *Business Insider UK*. Retrieved from https://www.businessinsider.com/
Marchiano, L. (2018, March 15). Outbreak: The explosion of transgender teens. *Mercatornet*. Retrieved from https://www.mercatornet.com/
Marsden P. (1998). Memetics and Social Contagion: Two Sides of the Same Coin? *Journal of Memetics: Evolutionary Models of Information Transmission, 2*(2), 171-185.
Marshall, T.A. (2018). *Rapid Onset Gender Dysphoria* [video file]. Retrieved from https://vimeo.com/282164120
Ofcom (2017). *Box Set Britain: UK's TV and online habits revealed*. Retrieved from https://www.ofcom.org.uk/about-ofcom/latest/media/media-releases/2017/box-set-britain-tv-online-habits
Raun, T. (2010). Screen-births: Exploring the transformative potential in trans video blogs on YouTube. *Graduate Journal of Social Science, 7*(2), 113-130.
—. (2016). *Out Online: Trans Self-Representation and Community Building on YouTube*. London: Routledge.
Rayner, G. (2018, September 16). Minister orders inquiry into 4,000 per cent rise in children wanting to change sex. *The Telegraph*. Retrieved from https://www.telegraph.co.uk/
Steinmetz, K. (2014, May 29). The Transgender Tipping Point. *TIME*. Retrieved from http://time.com/
Twenge, J. (2017, September). Have Smartphones Destroyed a Generation? *The Atlantic*. Retrieved from https://www.theatlantic.com/

Videography

Jace Rider. (2018). *Medical Transition Saved My Life - FTM Transgender*. Retrieved from https://www.youtube.com/watch?v=xGwWuqtnARE
JakeFtMagic. (2018). *TOP SURGERY MADE MY DYSPHORIA WORSE*. Retrieved from https://www.youtube.com/watch?v=SoF6mdBXF9Q
Jammidodger. (2017). *Trans Guy: Transitioning Saved My Life*. Retrieved from https://www.youtube.com/watch?v=R9WxSH1L9io

—. (2018). *The Dark Side of Testosterone*. Retrieved from https://www.youtube.com/watch?v=gFFgS-zLw98

Jett Taylor. (2018). *FTM - Coming to Terms with Yourself - Am I Trans?* Retrieved from https://www.youtube.com/watch?v=Dm6-O7iXJLI

Lonelygirl15. (2006). Retrieved from https://www.youtube.com/user/lonelygirl15/videos

MilesChronicles. (2018). *Watch this if you have questioned your gender*. Retrieved from https://www.youtube.com/watch?v=40IC0BU0Qto

Sam Collins. (2016). *DEAR PARENTS OF TRANS YOUTH*. Retrieved from https://www.youtube.com/watch?v=gIgs3YrQGaU

ThatGuyOli. (2017). *First Gendercare Appointment | Mission To Manliness*. Retrieved from https://www.youtube.com/watch?v=F6pPbFTAK_c

TheRealAlexBertie. (2014). *TO THE GP! 01*. Retrieved from https://www.youtube.com/watch?v=_sMp_lFgZJw&list=PLdPXV-RSgSKIhFP7RGqRCJ0lZWQ6e73cF

—. (2018). *Dealing with Depression & Starting Medication*. Retrieved from https://www.youtube.com/watch?v=1rFLPisBQ1w

Ty Turner. (2015). *How To Tell If You Are Transgender*. Retrieved from https://www.youtube.com/watch?v=f1rT7xOumO4

YouTube. (2015). *Top 4 Transgender YouTubers - The YouTube Buzz*. Retrieved from https://www.youtube.com/watch?v=ogjRiGrrieM

CHAPTER THIRTEEN

QUEERING THE CURRICULUM: CREATING GENDERED SUBJECTIVITY IN RESOURCES FOR SCHOOLS

STEPHANIE DAVIES-ARAI AND SUSAN MATTHEWS

Writing in *The Guardian* in 2017, Owen Jones claimed that 'Today's media-driven moral panic over trans people and their rights seems like history repeating itself' (Jones, 2017). To Jones, the issue is simple: concern about rising levels of gender dysphoria in children and young people replays events of the 1980s when social hostility towards homosexuality grew in the wake of the HIV/AIDS epidemic. 'Transphobia – Can we just skip it?', a video released by the left Labour pressure group Momentum in 2018, takes the same line, asking 'Is anyone else getting a sense of Deja vu?'

But history is not repeating itself. The issues raised by gender identity ideology in material for schools are not the same as those which led in 1988 to Section 28, an amendment to the Local Government Act which prevented 'the teaching in any maintained school of the acceptability of homosexuality as a pretended family relationship'. In 1986, the *Daily Mail* amongst other papers reported that a Danish children's book called *Jenny Lives with Eric and Martin*, about a girl parented by a gay couple, was available in ILEA schools. The issue gained political traction, and in 1987 Margaret Thatcher announced to the Conservative Party Conference that children were being taught that they have an inalienable right to be gay. It was only after the repeal of Section 28 in Scotland in 2000 and the rest of the UK in 2003 that LGB organisations were able to support young people in school.

The context in 2019 is different: in the UK, 'the number of children referred to GIDS [the nationally commissioned gender identity service for children] each week has risen by 2,500 per cent since 2009-10' (Channel 4

News, 2017). In 2015-16, 'patient numbers doubled compared to the previous year'. Since the Gender Recognition Act (GRA) was passed in the UK Parliament in 2004, support groups such as the Gender Identity Research and Education Society (GIRES, founded in 1997), Educate and Celebrate (founded by teacher Elly Barnes around 2005), Gendered Intelligence (set up in 2008) and Stonewall (which began to campaign on trans issues in 2015) have successfully lobbied to be the main providers of training on gender diversity for teachers and students. These organisations provide or disseminate picture books, story books and lesson plans for use in schools. GIRES offers the self-published *Being Me in Penguin Land* series together with lesson plans. Educate and Celebrate offers three 'Book Collections', priced at £200 each for early years, primary and secondary schools. Stonewall's education resources include 'Best practice, toolkits and resources'. Gendered Intelligence provides a resource called 'Knowledge is Power' and links to Educate and Celebrate and other providers. From 2018, teachers have also been able to consult the Transgender Trend school resource pack, described by Stonewall as 'a deeply damaging document, packed with factually inaccurate content'. Despite this criticism, Stephanie Davies Arai (founder of Transgender Trend and main author of the resource pack) was shortlisted for the 2018 John Maddox Prize which 'recognises the work of individuals who promote sound science and evidence on a matter of public interest, facing difficulty or hostility in doing so' (Sense about Science, n.d.).

This chapter will offer a close analysis of books and lesson plans disseminated for school use by trans support charities to argue that the messages transmitted to teachers, parents and children misrepresent science, ignore child development, fail child safeguarding and conflict with the law. These books foster homophobia, reinforce gender stereotypes and threaten a child's bodily integrity. It will argue that the current spate of books on gender identity does *not* replay the Section 28 controversy and that transsexuality 'is not *analogous* to homosexuality— either as a sexuality, as an object of study, or as an experience of being a sexual minority' (Hausman, 1995, p.3). Books and teaching materials provided to schools by trans-affirmative organisations endorse and may encourage a desire for body modification which is not part of homosexuality. In doing so, these resources may contribute to rising admissions to gender identity clinics for children.

Gender identity picture books

As Jones points out, *Jenny Lives with Eric and Martin* was not actually being used in schools in 1986: only 'a single copy had been purchased for the use of teachers' (Jones, 2017). By contrast, the *Being Me in Penguin Land* books for three to six year olds, offered by GIRES in 'trans boy', 'trans girl' and 'non-binary' versions, are available to teachers on the GIRES website and as 'beautifully presented glossy-cover paperbacks' (with the first copy free). Supposedly offered as a means to enable a trans child to reveal an inner truth, the books instead offer an adult template for a child's understanding of identity and an adult template for the process of communicating that identity to others. Each book begins with the parent penguin telling the infant, 'We can't always tell if you're a boy or you're a girl'. Human parents can reliably tell the sex of their child (except in the very rare case of a disorder of sex development where they will be offered professional help) but a child aged six may be too young to understand biological sex difference. In each version, the parents ask the infant penguin to reveal their sex: 'So tell us when you're ready [...]. We'll love you still, don't worry'. This question is repeated on the next page: 'So tell us when you're ready—there's no hurry. We'll love you still—don't worry'. Repetition suggests to the child that there *is* something to reveal— why else would the parents keep asking? This anxiety is endorsed by the narrator in the fourth slide who suggests, despite the parents' reassurance, that 'It can be easier to tell a friend'. This slide also supplies the words that the child will need to confide to their friend: 'I'm not a girl, cos I'm a boy'.

Omitting the moment at which the parents are told, the book skips to the narrator's statement: 'the family has listened and decided that you're right.' The child is renamed from Polly to Tom in the 'trans boy' version, whilst the 'trans girl' version reads: 'You know you're [sic] name is Sally, like you know you're black and white'. In the non-binary edition, the child (penguin) announces: 'I'm not a boy or a girl but somewhere in between. I'm not like all the others, I just don't feel the same'. This formulation requires the child reader to identity their difference from others and to do so on the basis of an ability to know how others feel. Appropriately, perhaps, the non-binary child gets the non-human name of 'Blur', thus revealing the extent to which the category of non-binary depends on social gender stereotypes. The target child reader could not identify the logical impossibility in this proposition. The *Penguin Land* books end with the parents throwing a party to celebrate the child's social transition, complete with balloons spelling the word 'Welcome'.

We know that children repeat behaviour that wins attention. The Transgender Trend *Resource Pack for Schools* suggests that 'Publicly 'celebrating' a transgender child as 'brave and courageous' can have unintended consequences. Schools should maintain a neutral stance of 'kind acceptance' (p.9). The GIRES series provides a narrow focus on gender categories coupled with a failure to encourage children to develop a language for the emotions. It also fails to recognise (as the Tavistock GIDS website reminds us) that:

> Children go through various stages of 'magical thinking,' during which they can get confused between reality versus fantasy, until at least middle childhood, and sometimes this makes it hard to know how much a younger child fully grasps about what they are saying or understands about their own gender.

The *Penguin Land* series is presented as a means to convey progressive attitudes to children too young to be contaminated by societal prejudice. Instead, animal characters are used for a didactic purpose that ignores the developmental stage of the child reader.

Social transition

The *Penguin Land* books advocate adult validation of a child's gender identity through the use of the pronouns and social gender stereotypes associated with the opposite sex. This is 'social transition' and requires every other child to deny their understanding of biological sex and to change their language accordingly. In *Can I Tell you About Gender Diversity?* (Atkinson, 2016), supplied by Educate and Celebrate as 'an ideal way to start conversations about gender diversity in the classroom or at home and suitable for those working in professional services and settings', twelve year old Kit explains that: 'When I asked people to start calling me Kit, and using he/him pronouns, this is called social transition' (p.17). Educate and Celebrate's Early Years Education Book Collection includes *Introducing Teddy: A gentle story about gender and friendship* (Walton, 2016). Here the switch from 'boy' to 'girl' takes place when 'Thomas' transforms his bow tie into a hair bow.

Adult validation of a child's understanding of gender is itself a powerful factor in the persistence of that identity. Speaking in 2018, Polly Carmichael, Director of the NHS gender identity development service, noted that an increasing number of young children referred to the service arrived having already made a social transition and were 'living in stealth' (Carmichael, 2018a). The Amsterdam gender clinic has recorded a similar increase: 'between 2000 and 2004, out of 121 pre-pubertal children, 3.3% had socially transitioned [...] when they were referred, and 19% were living in the preferred gender role in clothing style and hairstyle but did not announce that they wanted a change in name and pronoun' (Steensma et al, 2011). Between '2005 and 2009, these percentages increased to 8.9% and 33.3% respectively'. The 2011 study shows that social transition is not necessarily benign: 'two girls who had transitioned when they were in elementary school [...] struggled with the desire to return to their original gender role. Fear of teasing and feeling ashamed resulted in a prolonged period of stress'. GIDS cites the 2011 study on the Evidence Base on its website and recommends 'keeping options open and not having any preconceived ideas as [to] the longer-term outcome' (GIDS, 2019).

'Be who you really are'

'To thy own self be true': the advice given by Polonius to his son in Shakespeare's *Hamlet* derives from a tradition going back at least to the Delphic oracle. But transgender discourse offers a new slant on this age-old injunction. As Hausman points out: 'Endocrinology offered human subjects the fantasy of self-knowledge through chemical regulation and analysis. Transsexual discourses picked up this thread through the demand to change sex to match one's 'true sex' or 'true self' (Hausman, 1995, p.47). In gender identity books for primary school children, the imperative to 'be who you really are' is interpreted in terms of gender identity, ignoring both the reality of bodies (which may be judged to be wrong) and the myriad components of personality. In *Introducing Teddy: A story about being yourself*, the gender exploring teddy bear character Thomas laments 'I need to be myself'. The *Penguin Land* series is entitled 'Being Me'. In *Are You a Boy or Are You a Girl?* (Savage and Fisher, 2017), Tiny's triumphant last line is 'I am me!' When Jazz is allowed to wear 'girl clothes' to school in *I am Jazz* (Herthel and Jennings, 2015), we learn that 'being JAZZ felt much more like being ME'.

In transgender children's books, acting is often opposed to authenticity. *Jamie* (Pike, 2015) ranks characters according to the test of gender authenticity, and those who fail 'to be who they really are' are summarily dismissed. The princess is attracted to a trans man called Jamie because Jamie is 'not wearing a disguise'. Acting features in trans children's books as a means to uncover an authentic self rather than as a place for exploration: in both *George* (Gino, 2015) and *Gracefully Grayson* (Polonsky, 2014), acting a female part in the school play reveals the true female self of the central character. In *George*, the narrative is propelled by the teacher's refusal to consider George for the central role in a staging of *Charlotte's Web* because he is a boy. Whilst schools, in these books, refuse to allow a child to act the other sex, children are portrayed as incapable of acting another identity. This rigid understanding of identity undermines any claim to challenge the status quo. By contrast, the Transgender Trend *Resource Pack for Schools* encourages schools to 'Allow boys to take the female part and girls to take the male part in plays and performances' (p.14). Acting can provide an opportunity to empathise with other identities.

At the same time, these books actively construct a gendered subjectivity through the narrative point of view. In *George* (Gino, 2015) the narrator's use of 'she' to refer to a ten-year-old boy creates a contradiction that the reader expects will be resolved. Right from the start, the omniscient narrator knows that George is a girl—an effect that would not be achieved via a first-person narrative. Most people think of themselves in the ungendered first person, and first-person narration as developed in the twentieth century stream of consciousness typically allows an escape from gender: it is no coincidence that Virginia Woolf believed in the androgyny of the self. Most relations with others take place using the ungendered second person: the 'I/Thou' relationship described by Martin Buber does not require the specification of gender. But the third-person narrative in *George* encourages the reader to notice the conflict between the masculine name and female pronouns: 'George had to steady herself awkwardly on one foot' (Gino, 2015, p.1). The choice of a third-person narrative in these books suggests the self-objectification involved in the adoption of cross-gender pronouns.

Sexual orientation

The Educate and Celebrate guide *How to Transform Your School into an LGBT+ Friendly Place* (Barnes and Carlile, 2018) promises 'to smash heteronormativity'. But the transgender books this organisation promotes encourage child readers to ignore or to redesignate signs of emerging same-sex attraction. *Jamie* (Pike, 2015) retells the Cinderella story as a transgender fable with Jamie as a gender non-conforming girl who is 'great at fixing things' but is bullied by two older brothers who demand that she fix a motorbike or mend a car for them. In addition, Jamie is expected to 'clean the house, prepare the meals and sew up all the holes in her brothers' stinky socks'. Her brothers have their own fairy godmother who gives them anything they wish for. But far from challenging sexist stereotypes, this revision of Cinderella suggests that becoming a boy is the only way to escape bullying and domestic drudgery. Jamie 'didn't feel right at all not in her body that is' because 'how she felt inside didn't match up with what she saw in the mirror'.

For a girl attracted to other girls, the message is that if you like having short hair and wearing trousers you are not a lesbian but a boy. The story of *Jamie* takes diversity and difference (gender non-conforming, same-sex attracted) and turns it into conformity (gender-conforming, heterosexual). Similarly, George might well be gay: 'George thought for a moment about kissing a boy, and the idea made her tingle' (Gino, 2015, p.33). The female pronoun in this sentence proactively denies a homosexual interpretation. In both *George* and *Gracefully Grayson*, the words 'gay' and 'girl' are used interchangeably as insults towards a boy who does not perform masculinity adequately. In *George,* we hear this from his classmates: 'Jeff smirked 'He's such a freaking girl anyway.' Jeff guffawed, and Rick laughed alongside him' (p.89). The conflation of femininity with homosexuality in a boy is strong; even after Grayson 'comes out' as a 'girl', his cousin cannot let go of the connection: ''Well, Grace, I think you'll make a perfect girl', Jack says. 'But what am I supposed to tell all my friends when they ask me why my cousin is totally gay?'' We know that extreme gender non-conformity in childhood is strongly correlated with later homosexuality.

A 2017 study revealed that boys who behaved in ways more typical of the opposite sex were eight times more likely to identify as homosexual at age 15 than those whose behaviour was more typical of their sex (Li, Kung, and Hines, 2017). Even so, the great majority of extremely gender non-conforming children grow up to be heterosexual. Rather than challenging homophobia or accepting gender non-conformity, transgender children's books reinforce the messages of the bullies. We know that the kind of sustained homophobic bullying which is experienced by both boys in *George* and *Gracefully Grayson* can affect the development of gender identity (De Lay et al, 2017). Any pre-gay boy reading these books will learn that transition to become a girl is a possible solution.

What are gender non-conforming children to make of this? At an age when peers are busy policing gendered behaviour, they are the children who are most likely to experience teasing, bullying and isolation from peer groups. No alternative way to be your 'true self' is suggested in these books other than switching from one sex to the other. No boy who loves princesses and ballet is celebrated for being his true authentic self as a boy who loves princesses and ballet. For a child like this, the idea that he may really be a girl will reinforce what he is no doubt already hearing from classmates, because this is the level of understanding of the primary-aged child. That is why the Transgender Trend *Resource Pack for Schools* specifically addresses the danger from gender identity ideology to children who are same-sex attracted with the advice: 'When teaching children about sexual orientation, clarify that gay and lesbian people are same-sex attracted and not 'same gender' attracted, i.e. a gay man is sexually attracted to males and a lesbian is sexually attracted to females' (p.15). The *Resource Pack for Schools* includes a statement from the Lesbian Rights Alliance which warns that 'young lesbians in schools who do not conform to feminine stereotypes (sometimes also labelled as 'butch') are being bullied, stigmatised, isolated and pressurised to socially transition since being a trans boy is now regarded as a more positive and fashionable identity' (p.21). The *Resource Pack for Schools* encourages schools 'to find positive role models who are 'butch' or 'dyke' lesbians or 'effeminate' gay men to come in to talk to secondary school students' (p.14). This is not a rerun of the 1980s.

Gender stereotypes and the creation of 'cis'

Trans school materials often associate gender non-conformity with a transgender outcome. In *Are You a Boy or Are You a Girl?* (Savage and Fisher, 2017), Tiny, a child of unspecified sex, dresses up as a butterfly but also enjoys football. Tiny's dad is a bus driver who cooks when he gets home. Firefighters visit the school and the kids discover that 'there is a lady driving the fire engine'. There seems no need for Tiny to change whatever sex they are to access a full range of social gender options. So far, the message matches that of the Transgender Trend *Resource Pack for Schools* which advises to 'Invite in adults to speak who defy gendered expectations in their professions, e.g. a female firefighter, a male nurse' (p.14). Yet the book slides from a rejection of gender stereotypes into a denial of sexual difference and ends with a series of questions: 'How are girls and boys different?', 'Do you think that Tiny is a boy or a girl?', 'Does it matter if Tiny is a boy or a girl?' The child reader is denied the ability to recognise the biological reality of sex and encouraged to associate gender non-conformity with transgender. The book ends with the contact details of GIRES and Mermaids, 'the only UK charity providing support for children and gender identity issues and their families'.

Books for children approaching puberty promoted by transgender lobby organisations often reinforce gender stereotypes. George's femininity, for instance, is asserted through a fascination with stereotypically gendered teen magazines whose covers 'promised HOW TO HAVE PERFECT SKIN, TWELVE FRESH SUMMER HAIRCUTS'

(Gino, 2015, p.2). Kit, in *Can I Tell You About Gender Diversity?*, explains the need to become a boy thus: 'I didn't like playing with dolls, or wearing dresses, and I hated having long hair' (Atkinson, 2016, p.7). Female readers who are not stereotypically feminine must question whether they are really girls. Where we might hope to see books for young readers that would challenge stereotypes, these books embrace stereotypes as proof of gender identity. Yet the desire to conform to gender stereotypes (of the other sex) is presented as a challenge to gender stereotypes. Stonewall even offers *I am Jazz* (2015), with its deeply stereotypical presentation of cultural femininity, under the category 'Challenging Gender Norms and Stereotypes' in its 'Primary School Book list'. 'For as long as I can remember, my favourite color has been pink', announces Jazz, as if this fact explained a desire for body alteration.

Whilst trans-identifying children long to adopt the stereotypes of the other sex, the 'cisgender' child is constructed in this literature entirely from negative cultural stereotypes. The representation of boys who are not trans is overwhelmingly negative. In *George,* the trans child cries at the death of Charlotte, whereas the 'cis' Jeff is only interested in killing:

> 'I'd step on her. Crush her under my foot like the freak she is. Freaky spider. Stupid freaky spider.' Jeff began to sing an unformed tune. '*Stupid, freaky spider. I'm gonna step on you because it's what you deserve, you stupid, freaky spi-i-der. I'm glad you diiiiiiiiieeed*'. (Gino, 2015, p.114)

Jeff also calls George a freak: 'You're such a freak. You're a freak. Freak. Freak' (p.118). By contrast, George's friend Kelly teaches him to braid hair, apply make-up and wear a skirt:

> She dashed to the closet and pulled out a flared skirt of purple swirls and rummaged through a drawer for a hot-pink tank top. She laid the clothing in George's hands. The top was soft, softer than any boys' shirt she had ever worn. And she had never held a skirt in her hands like this before. Together they felt magical. (p.180)

The reader is offered a choice between death-loving masculinity and nurturing, sensitive and tactile femininity. Although George is told that boys played female parts in Shakespeare's time (p.33), that men can explore femininity (p.10) and that he may be gay (p.138), these ideas simply confuse a ten-year-old with an intense dislike of the word 'man' in a story which offers few if any positive male role models.

Imposing the label 'cis' makes it clear that gender ideology requires the redefinition of everyone, including oneself, as a 'gender identity'. 'Cisgender' identity is revealed as a willingness to conform to sexist stereotypes. The transgender child is framed as the only one who does not conform, and being transgender becomes the only way of expressing non-conformity. What child approaching the teenage years would want to be in the boring conformity camp?

Persistence, desistance and the arrival of puberty

Pre-pubertal gender dysphoric children are far more likely to grow out of their cross-sex identification than to retain these feelings through adolescence (Cantor, 2017). Yet transgender books offered to school children misrepresent the evidence by suggesting that gender identity is unlikely to change: 'Tonight Jamie had felt right for the first time and he knew this would be how things would stay' (Pike, 2015). This message is supported by the GIRES presentation for children aged 11-14 on 'The Gender Question' (Key Stage 3), which asserts as scientific fact the theory that gender identity is innate and a product of the gendered brain. '[J]ust to reinforce the point', a slide announces, 'it's the brain that tells you whether you are a boy or a girl.' The illustration shows two brains, one with a pink spot and one with a blue spot. The slide continues: 'The brains of trans people are 'hard-wired' slightly differently from the majority of the population. These differences have been found in a number of studies on the brains of trans people, compared with the brains of non-trans people'.

GIDS does not endorse this theory. To the question 'Why does my child feel this way?', the GIDS website replies:

> The honest answer is we do not know exactly why any particular child or young person develops cross-gender or gender variant feelings or behaviour. Nor can we pinpoint why any one individual might develop a transgender identity. For each individual there are likely to be a variety of different factors that come together to shape them as a person, including biological, social and psychological factors, as well as their experiences as they are growing up. (GIDS, n.d.)

GIDS mentions that '[s]ome people think that young people are 'born this way'' and links to the GIRES website but also refers parents to Cordelia Fine's *Delusions of Gender*. In a 2014 article, GIDS clinician Bernadette Wren recognises that '[m]any young transgender people bolster their identity claims with a belief that such evidence will soon be available to show they literally have 'male brains in female bodies', or vice versa.' (Wren, 2014) These are beliefs and hopes, according to Wren.

Rather than seeking out the balanced evidence available on the GIDS website, *Can I Tell You About Gender Diversity?* promotes YouTube transition videos which already have hundreds of thousands of predominantly teenage girl followers and have been identified as a factor in social contagion (Littman, 2018). According to Kit: 'Something that has really helped me has been looking at all the videos that people put up on YouTube. There are lots of boys and girls and non-binary people who make videos every day talking about what their bodies mean to them, and how they feel about their gender' (Atkinson, 2016, p.23). Speaking in 2018, Dr Polly Carmichael, director of GIDS, said she had been

> so shocked by some of the things that are swilling around the internet that young people have access to. There are numerous groups on Reddit and Tumblr that many of the young people attending our service are going on to. [...] Maybe it's also the dissing of expertise, in a way, so there's a feeling that this is about who I am so what does anyone else know. (Carmichael, 2018b)

Presenting twelve-year-old Kit as a source of authority only reinforces this message. In trans books for school children, friends and YouTube videos are suggested as the best way to find out about transitioning. George's friend Amelia turns to the internet to discover that 'you could take girl hormones that would change your body, and you could get a bunch of different surgeries if you wanted them and had the money. This was called *transitioning*. You could even start before you were eighteen with pills called androgen blockers that stopped the boy hormones already inside you from turning your body into a man's' (Gino, 2015, p.47). Body changes at puberty can produce severe anxiety, especially for girls. The message that common feelings of body hatred at puberty are a sign of being transgender is therefore particularly dangerous, but this is just what Kit suggests: 'My friend Amy, who I met at the clinic, didn't really start to feel dysphoric until she was 12 and her body started changing in a way that made her feel really bad about herself. That happens for people a lot, when your body changes and you realise you don't want it to change in a certain way' (Atkinson, 2016, p.25).

Medical transition is portrayed in these books as a simple process to prevent unhappiness, on the same level of seriousness as changing clothes and pronouns. 'Medical transition for people my age is usually hormone blockers. These stop me from going through female puberty, and stop my body developing in ways that make me unhappy' is pronounced by Kit as matter-of-factly as 'I wear boy's clothes, and I have a boy's hair cut' (Atkinson, 2016, p.18).

Medical interventions

Much of the information given about medical transition in the books offered to schools by trans support charities is inaccurate and potential harms are ignored, glossed over or falsified. The twelve-year-old Kit explains to the child reader that 'When I'm 18 I'll be allowed to go onto hormone replacement therapy—this means that I'll start to develop facial hair, and my body will go through boy puberty' (Atkinson, 2016, p.19). But cross-sex hormones do not allow the body to go through puberty: 'Oestrogens and testosterone induce masculine or feminine physical characteristics and SHOULD ONLY BE taken in the context of medical supervision to monitor risks (e.g., polycythaemia in transgender males, venous thromboembolism in transgender females)' (Heneghan & Jefferson, 2019). These children's books misrepresent the state of medical knowledge, falsely suggesting that transition is an accepted and safe medical process. Heneghan & Jefferson conclude:

> Treatments for under 18 gender dysphoric children and adolescents remain largely experimental. There are a large number of unanswered questions that include the age at start, reversibility; adverse events, long term effects on mental health, quality of life, bone mineral density, osteoporosis in later life and cognition. We wonder whether off label use is appropriate and justified for drugs such as spironolactone which can cause substantial harms, including death. We are also ignorant of the long-term safety profiles of the different GAH regimens. The current evidence base does not support informed decision making and safe practice. (Heneghan, 2019)

It is therefore both misleading and unethical to promote a book for children in which the protagonist claims that 'the best thing about hormone blockers is that if I change my mind then they won't hurt my body' (Atkinson, 2016, p.28). There is no evidence at the time of writing on whether puberty blockers are fully reversible when taken during normal puberty, or if they cause permanent harm (Jarrett, 2018). The presentation of unchecked claims about drug-based interventions through books in schools is an irresponsible minimisation of invasive and experimental treatment, which seems only designed to sell it to young people. Double mastectomy is suggested to girls at puberty who hate their developing breasts: 'My friend Tobi doesn't think that they want to go through hormones but might have top surgery so that they don't have breasts, because they just don't feel comfortable in their body as it is, but they don't feel binary about gender' (Atkinson, 2016, p.28).

Schools guidance, accuracy and the law

The messages promoted to young children through transgender storybooks are echoed in Stonewall's guidance for teachers, written with Gendered Intelligence (Stonewall, 2015) and repeated in the guidance from Educate and Celebrate and GIRES (GIRES, 2014). Stonewall's 54-page *Introduction to Supporting LGBT Young People* school guide includes nine pages on transgender pupils alone, and the message of the three case studies is that teachers should actively push children towards transition. One case study shows a teacher helping a child to persuade their parent or carer to support their transition:

> 'Yes, but they're not supportive—my mum/dad/carer won't speak to me about it.'
> 'Well we can try and help you with that. Is there anyone else in your family who you can talk to? There are organisations that can help—I can give you their details.'

The teacher here helps the child to access an organisation behind their parents' backs. In the next case study, a young person who wants more time to think is offered further information and encouraged to come back to follow the issue up:

> 'I don't know what I want to do—I need to think some more. I just know that I don't feel happy and right the way I am at the moment.'
> 'That's okay and it's good to take some time to think things over. I will point you in the direction of some information that might be useful. Why don't you come and talk to me once you've had a look?'

In the final case study, the teacher is advised to insist that the school will do everything within its power to make transition possible, despite the pupil's doubts:

> 'I think I want to take steps to live as the gender I know I am but I'm worried about how it will work at school.'
> 'The school is here to make sure things feel right for you. We can arrange a time to sit down and talk through all the options and different ways a transition might work at school. What do you think? There are lots of people who have transitioned at school—it is possible!'

The Stonewall account of transition echoes that of the twelve-year-old Kit, because Kit is really C. J. Atkinson, an adult academic and activist. Both describe the move from social transition to medical transition as if medically altering your body is on a par with changing your wardrobe. 'A trans young person may feel unhappy or distressed about living with a body they don't feel reflects their gender identity. Some young people choose to make changes to their body through hormone treatment, though this can involve waiting a long time. Schools can help by ensuring that young people know how to access support services, can talk to others and learn about self-esteem and body confidence in PSHE.' The idea that your body is 'wrong' and needs fixing to match your personality is incompatible with lessons on body confidence and acceptance. The hypocrisy in teaching children this belief and framing it as encouragement to a young person to feel positive about themselves is exposed in Stonewall's guidance: 'When a young person comes out it is important to reinforce that they can be themselves and encourage them to feel positive about who they are.' This echoes the 'being yourself' narrative, which has been sold to children as the only way to understand their feelings.

Not only does Stonewall direct teachers to encourage and facilitate pupils' transition, they advise doing so behind parents' backs: 'Not all young people will want their parents/carers to know they are lesbian, gay, bisexual or trans, and for staff to discuss this with parents/carers without the young person's consent would be a breach of confidentiality'. Here, Stonewall breaks the first rule of safeguarding, which is that a teacher should never promise confidentiality to a child. This rule is there for good reason; no member of staff can unilaterally decide on the best course of action for any one child. Stonewall suggests here that a girl announcing she is a boy is no different to a girl who comes out as a lesbian.

Increasing numbers of young people who identify as transgender have pre-existing mental health problems, past trauma or troubled backgrounds. In one recent study, 10% had suffered past sexual abuse (Bechard et al, 2017). 35% of children referred to the Tavistock clinic exhibit moderate to severe autistic traits (Butler et al, 2018). It is crucial that schools offer proper support to the most vulnerable young people and all underlying factors must be considered as part of a school's safeguarding duties towards each individual child. Advice which forces teachers to collude in the affirmation approach towards young people with gender dysphoria makes this impossible.

Stonewall insists 'there are absolutely no issues under child protection or safeguarding laws specific to trans-identified and gender variant young people'. Recasting the gender dysphoric child as a political category prevents teachers from identifying any underlying circumstances by framing any other response but 'affirmation' as transphobic. The reason for teaching children from the youngest age that it does not matter if you are a boy or a girl and nobody has the right to question anyone else is revealed when we get to the guidance on single-sex facilities. All three organisations misrepresent equality law by asserting that a 'trans' pupil has the right to access all sex-segregated facilities, sports and overnight accommodation. Access to single-sex spaces is crucial to support the ideology that it is your 'gender identity' which makes you a boy or a girl and not your sex.

Insisting on mixed-sex toilets, changing rooms and sleeping accommodation denies all children their right to privacy from the opposite sex, puts girls at risk and undermines basic safeguarding. Policies based on 'gender identity' erase boundaries between the sexes along with the principle of consent. Denying biological sex also means denying same-sex attraction, which Stonewall redefines as 'attraction towards someone of the same gender'. Educate and Celebrate clarifies what that actually means in their definition of 'lesbian' as a 'person who identifies as female who is romantically, emotionally and/or physically attracted to another person who identifies as female'. The substitution of 'identity' for reality is complete. This redefinition allows heterosexual men to identify themselves as 'lesbians,', thus destroying lesbian-only spaces and coercing young lesbians to accept fully-intact males as sexual partners.

Conclusion

According to Owen Jones, 'History is a savage judge of those who resisted the onward march of gay rights. I doubt it will be less damning of those who bitterly fight trans rights' (Jones, 2017). But the didactic children's literature currently on offer to schools imposes adult messages on their child readers that are not progressive, that occlude gay and lesbian identities, that encourage a child to dissociate from their body and that misrepresent current thinking on gender dysphoria. These books are less progressive than a poem for children written over two hundred years ago. William Blake's *Songs of Innocence* (1798) have long been seen as a reaction against the moralism and didacticism of children's literature of his own time. His poem 'Infant Joy' might be appropriate for a child aged three to six (now offered a *Penguin Land* book):

> I have no name
> I am but two days old.—
> What shall I call thee?
> I happy am
> Joy is my name,—
> Sweet joy befall thee!

Pretty joy!
Sweet joy but two days old,
Sweet joy I call thee;
Thou dost smile.
I sing the while
Sweet joy befall thee.

Blake's poem presents the voices of a child and carer in dialogue. But it also admits its own artifice. The reader knows that the poem cannot be the words of a two-day old child: the word 'infant' derives from the Old French 'enfant' or from the Latin 'infant', meaning 'unable to speak'. Blake's poem reminds us that children's literature literally puts words into the mouths of children, teaching them both to speak and to shape their experience according to adult concepts. This is the case with all children's literature, as Jacqueline Rose explained, but it is a truth that is forgotten or concealed by didactic children's literature of our own time (Rose, 1984). The voice of the twelve year old Kit is generated by C. J. Atkinson, adult activist and academic, and it offers children a range of adult identities. Blake's poem, by contrast, does not gender the child but represents feelings which are echoed and validated by the adult. This poem is both more honest and more open than the children's books supplied as progressive by trans support groups to today's school children.

By allowing trans and queer activists to dictate policy, teachers are being led to support a clinically-contested, controversial and experimental 'affirmation' approach to gender dysphoria. Under the guise of 'LGBT inclusion', all children are being conditioned to believe in the concept of innate gender identity and to accept without question that some boys are in reality girls and vice versa. Girls are unaware that they are being groomed into a belief in gender which allows them no sexual boundaries. Through the denial of biological sex, they are disenfranchised by an ideology which takes away the protection of girls as a sex under the Equality Act 2010, effectively erasing the female sex as a distinct class in culture and law. Transgender resources in schools are ensuring that this generation will grow up thoroughly conditioned to support one side of the ideological and political debate. If we can convince girls from an early age to believe that 'woman' is an identity they can choose and that some women have penises, we prevent the development of a feminist political consciousness, therefore conveniently erasing any future feminist opposition to gender ideology of the kind we are seeing so violently suppressed today.

References

Atkinson, C.J. (2016). *Can I Tell You About Gender Diversity?* London and Philadelphia: Jessica Kingsley.

Barnes, E., and Carlile, A. (2018). *How to Transform Your School into an LGBT+ Friendly Place*. London and Philadelphia: Jessica Kingsley.

Bechard, M., Vanderlaan, D.P., Wood, H., Wasserman, L., and Zucker, K. (2017). Psychosocial and Psychological Vulnerability in Adolescents with Gender Dysphoria: A "Proof of Principle" Study. *Journal of Sex and Marital Therapy, 43*(7), 678-688.

Blake, W. (n.d.). Infant Joy. *Poetry Foundation.* Retrieved from https://www.poetryfoundation.org/poems/43665/infant-joy

Butler, G., De Graaf, N., Wren, B., and Carmichael, P. (2018). Assessment and support of adolescents with gender dysphoria. *Archives of Disease in Childhood, 103*, 631-636.

Cantor, J. (2016, January 11). Do trans kids stay trans when they grow up? *Sexology Today.* Retrieved from http://www.sexologytoday.org/

Carmichael, P. (2018a). *Genetic perspective - differences and similarities between gender dysphoria and sex development disorders.* Paper presented at 'Science of Gender: Evidence for what influences gender development and gender dysphoria and what are the respective influences of nature and nurture', 18-19 October, Tavistock Centre, London.

—. (2018b). *The Association for Children and Adolescent Mental Health Emanuel Miller Lecture.* Paper presented at the Emanuel Miller Memorial Lecture & National Conference, March 16. Available at https://twitter.com/acamh/status/974676428972220417

Channel 4 News (2017). Fact Check Q & A: How many children are going to gender identity clinics in the UK? *Channel 4.* Retrieved from https://www.channel4.com/news/factcheck/factcheck-qa-how-many-children-are-going-to-gender-identity-clinics-in-the-uk

De Lay, D., Martin, C.L., Cook, R.E. and Hanish, L.D. (2017). The Influence of Peers During Adolescence: Does Homophobic Name-Calling by Peers Change Gender Identity? *Journal of Youth and Adolescence, 47*(3), 636-649.

Educate and Celebrate (2016). *Primary book collection - Early years book collection.* Retrieved from http://www.educateandcelebrate.org/wp-content/uploads/2016/09/Early-Years-Book-Collection-2016.pdf

Gino, A. (2015). *George.* New York: Scholastic.

GIDS (2019). *Evidence Base.* Retrieved from http://gids.nhs.uk/evidence-base GIRES (2014). *Transition of a Pupil in School.* Retrieved from https://www.gires.org.uk/wp-content/uploads/2014/08/Transition-of-a-Pupil-in-School.pdf

GIRES (2015). *Classroom lesson plans*. Retrieved from https://www.gires.org.uk/classroom-lesson-plans/

Hausman, B.L. (1995). *Changing Sex: Transsexualism, Technology and the Idea of Gender*. Durham and London: Duke University Press.

Heneghan, C. and Jefferson, T. (2019). Gender Affirming Hormone in Children and Adolescents. *British Medical Journal Evidence Based Medicine Spotlight*. Retrieved from https://blogs.bmj.com/bmjebmspotlight/2019/02/25/gender-affirming-hormone-in-children-and-adolescents-evidence-review/

Herthel, J. and Jennings, J. (2015). *I Am Jazz*. New York: Dial Books.

Jarrett, C. (2018). Systematic review finds 'qualified support' for hormonal treatments for gender dysphoria in youth. *British Psychological Society Research Digest*. Retrieved from https://digest.bps.org.uk/2018/07/23/systematic-review-puberty-suppressing-drugs-do-not-alleviate-gender-dysphoria/

Jones, O. (2017). Anti-trans zealots, know this: history will judge you. *The Guardian*. Retrieved from https://www.theguardian.com/

Littman, L. (2018). Rapid-onset gender dysphoria in adolescents and young adults: a study of parental reports. *PLOS ONE, 13*(8), e0202330.

Li, G., Kung, K.T.F. and Hines, M. (2017). Childhood Gender-Typed Behavior and Adolescent Sexual Orientation: A Longitudinal Population-Based Study. *Developmental Psychology, 53*, 764-77.

Pike, O. (2015). *Jamie*. London: Oliver Pike.

Polonsky, A. (2014). *Gracefully Grayson*. White Plains, NY: Disney-Hyperion.

Rose, J. (1984). *The Case of Peter Pan or the Impossibility of Children's Fiction*. London: Palgrave Macmillan.

Savage, S. and Fisher, F. (2017). *Are You a Boy or Are You a Girl?* London and Philadelphia: Jessica Kingsley.

Sense about Science (n.d.). *The John Maddox Prize*. Retrieved from https://senseaboutscience.org/activities/the-john-maddox-prize/

Steensma, T. D., Biemond, R., de Boer, F., & Cohen-Kettenis, P. T. (2011). Desisting and persisting gender dysphoria after childhood: A Qualitative follow-up study. *Clinical Child Psychology & Psychiatry*, 16(4), 499–516. Retrieved from https://doi.org/10.1177/1359104510378303

Stonewall (n.d.). *Primary school book list*. Retrieved from http://www.stonewall.org.uk/sites/default/files/reading_list_primary_final_lo_res_v2.pdf

—. (2015). *An Introduction to Supporting LGBT Young People: A Guide for Schools*. Retrieved from https://www.stonewall.org.uk/system/files/an_introduction_to_supporting_lgbt_young_people_-_a_guide_for_schools_2015.pdf

Transgender Trend (2017). Hot topics in child health: a medical, political and ethical debate. *Transgender Trend*. Retrieved from https://www.transgendertrend.com/hot-topics-in-child-health-a-medical-political-and-ethical-debate/

Walton, J. (2016). *Introducing Teddy*. New York: Bloomsbury.

Wren, B. (2014). Thinking postmodern and practising in the enlightenment: Managing uncertainty in the treatment of children and adolescents. *Feminism & Psychology, 24*(2), 271-291.

CHAPTER FOURTEEN

GENDER GUIDES AND WORKBOOKS: UNDERSTANDING THE WORK OF A NEW DISCIPLINARY GENRE

SUSAN MATTHEWS

The gender journey, the gender revolution

In January 2017 a 'special single-topic issue' of *National Geographic* magazine announced a 'Gender Revolution' with a cover photograph of a nine-year-old trans girl, Avery Jackson. A downloadable 'Discussion Guide' accompanied the magazine to teach the basics of the new gender knowledge:

> This discussion guide for teachers and parents is not a discussion guide on sex or sexual orientation. Rather, used in conjunction with the magazine and film, it is a tool to help you understand the nature of gender and its ramifications as we work together toward a more inclusive and tolerant world. (National Geographic, 2017).

A two-hour documentary, 'Gender Revolution: A Journey with Katie Couric' (hashtag #GenderRevolution), aired the following month. In March 2018 the 'gender revolution' was presented in *The Guardian* by Google creative director Tea Uglow as an epistemic shift comparable to 'the Copernican revolution'. The gender revolution is now being disseminated to a popular audience via guides and workbooks. Sally Hines's *Is Gender Fluid? A Primer for the 21st Century* (2018) contains 'over 160 illustrations' and uses varied font sizes to guide the reader towards the key propositions: 'The larger the font size the more important the words are to the overall concept or argument.' If we can just 'set aside a couple of hours' we can get up to speed on ideas introduced by writers as difficult as Michel Foucault, Judith Butler and Thomas Laqueur. The index to Iantaffi and Barker's 2017 *How to Understand your Gender*, 'a practical guide for exploring who you are', lists references to Sara Ahmed, Judith Butler, Simone de Beauvoir, Michel Foucault and Luce Irigary. Guides and workbooks offer to translate heavyweight academic theory into forms that will change our lives: this is queer theory made flesh and come amongst us (see Bornstein 1998, 2013).

Paradoxically the gender revolution is presented both as new and as a return to age-old human freedoms. According to Tea Uglow: 'Trans identity is not new—it is only a surprise because we suppressed it for so long'. According to S. Bear Bergman: 'Trans and genderqueer and non-binary are not remotely new things. That they—that we—are now visible in greater number than during any time in recorded history is just the natural order, reasserting itself' (Iantaffi & Barker, 2017, p.12). According to Iantaffi and Barker: 'most liberation starts from a seed and we feel the seed of gender liberation has been planted for some time—it has blossomed in the past and it might indeed bloom again' (2017, p.85). Underlying these claims is the idea that the new gender freedoms are a return to an older, precolonial or precapitalist order. The repression of gender freedom, according to this narrative, was the work of colonization and Christianity:

> Christianity, as practiced in the time, had no room at all for anyone beyond the gender binary of male and female, even though prior to colonization many, many indigenous societies had neutral-to-positive words, social roles, and legal standards for people we might today name as transgender in some way. (Iantaffi & Barker, 2017, p.11)

It makes sense, then, that the *National Geographic*'s 'Gender Revolution' offers a tour through gender systems in non-western cultures: 'South Asia (where a third gender is called hijra), Nigeria (yan daudu), Mexico (muxe), Samoa (fa'afafine), Thailand (kathoey), Tonga (fakaleiti), and even the U.S., where third genders are found in Hawaii (mahu) and in some Native American peoples (two-spirit)'. As the US magazine disseminates the freedoms of non-western cultures to a global audience, gender freedom passes from countries formed by white colonial expansion (Canada, US, Australia and New Zealand) through the communication networks of global capitalism to readers around the world. This chapter asks to what extent the 'gender revolution' deserves to be recognized as a form of liberation and questions whether the 'gender journey' offers an escape from gender ideology.

Why do kids need workbooks?

Gender identity workbooks began to appear at the end of the twentieth century, but it was after 2014, the 'transgender tipping point' according to *Time Magazine*, that publishers began to offer gender identity workbooks for kids. The *Gender Quest Workbook: A Guide for Teens and Young Adults Exploring Gender Identity* (Testa, Coolhart & Peta, 2015) describes a journey with references to 'pathways', taking 'small steps' (p.vii), the 'exciting journey of self-discovery' (p.1). The workbook takes the place of a friend on this journey: 'If you have anyone else you trust to go through this book with you, great. [...] Otherwise, don't fret, we'll be by your side throughout the journey' (p.4). An Afterword by gender clinician Arlene Lev explains that 'All of life is truly a quest – a journey and explorative odyssey – to define, invent, discover, and create a unique self, one that reflects our inner experience' (p.167). Quest, journey, odyssey—metaphors familiar from the earliest epic, through *Pilgrim's Progress* to nineteenth century boys' adventures stories—are recruited to figure the gender journey. But rather than Huckleberry Finn's journey down the Mississippi, the twenty-first century adolescent can go on a 'Gender Quest' in the 'private space' of the workbook.

Gender identity workbooks are a new subdivision of the self-help industry that developed in the last quarter of the twentieth century as 'Americans turn to self-improvement literature for inspiration in times of despair, for specific advice on how to conduct their lives, and for reassurance in the face of enormous social, political, and economic changes' (McGee, 2005, p.17). McGee argues that 'as recently as the 1970s [...] 'self-help' referred not to individual self-improvement practices but to cooperative efforts for mutually improved conditions' (2005, p.18). In the self-help book, individual self-invention now overcomes the power of family, society, economy, and culture to define the self. The workbook, 'containing information on a subject and related exercises for students, such as questions to be answered or tasks to be carried out' (OED Online), is a format designed to instil lessons by active engagement. Learning may be packaged by the *Gender Quest Workbook* as 'fun and engaging activities and exercises' but this is still work. And in *The Gender Identity Workbook for Kids*, the task of individual self-invention becomes a duty required of children. *The Gender Identity Workbook for Kids* is aimed at seven-year-olds who are 'too old for the little kid picture books' (Storck, 2018, p.xii). In *Who Are You?*, a picture book for three to five-year-olds, the pursuit of gender identity is the task of '[e]very reader, from age 1 to 101' (Pessin-Whedbee, 2017). In *How to*

Understand Your Gender, anyone 'thinking about your own gender right now', 'finding out about gender for the first time', even professionals 'if they haven't yet had the opportunity to do reflective work on their own gender', were offered a space to work on their gender identity (Iantaffi & Barker, 2018, p.21).

The gender identity workbook for children and young people arrived on the publishing scene in the twenty-first century in the wake of 'puberty blockers' or GnRH agonists, which made it possible for children to avoid their natal puberty and the acquisition of secondary sex characteristics (breasts, Adam's apple, broken voice). These interventions, first used at the Amsterdam gender clinic in the late 1990s, were subsequently made available at clinics including GEMS in Boston from 2007 and the London GIDS for children and young people from 2011 onwards. What followed was an exponential rise in referrals of children and adolescents to gender identity clinics across the developed world. Julia Serano's 2016 introduction to the second edition of *Whipping Girl* notes that 'trans children [...] socially transitioning prior to adulthood [...] was still rare back when I was writing this book' (xxiv); in other words, the transgender child came to prominence in the years between 2007 and 2016. In 2018, Christine Burns dedicated *Trans Britain: Our Journey from the Shadows* to the transgender children's support charity Mermaids, lamenting the 'negative and irreversible bodily changes' (otherwise known as natal puberty) that adult transgender people have in the past had to endure (n.p). Burns looks forward to a world in which children, identified as transgender before puberty, develop into cosmetically passing adults. If 'developments in technology make new discursive situations possible' (Hausman, 1995, p.14), transgender kids can be seen as the product of a new medical technology.

Yet the workbooks represent transgender identity as innate, a product of the pre-social knowledge of gender freedom repressed by cultural gender expectations. The work of the workbooks is to occlude and to naturalize technology, creating a myth which 'radically erases the semiotic conditions of its own existence' (Hausman, 1995, p.ix). Julia Serano describes transgenderism as 'a profound, inexplicable, intrinsic self-knowing' which has nothing to do with social gender stereotypes (2007, 2016, p.151). Serano prefers the term 'unconscious sex' to 'gender identity', implying that the unconscious speaks directly to the transgender person. Access to 'unconscious sex' requires a kind of reprogramming which can strip away existing gender beliefs: 'If we are to have any hope of truly understanding gender, we have to find a way to undo our gender brainwashing' the teenage reader of the *Gender Quest Workbook* is told

(Testa, Coolhart & Peta, 2015, p.15). To recover a primal gender scene, a traumatic moment of entry into gender ideology, the gender traveller must ask: 'What are your earliest memories related to gender? When was the first time you understood how your gender would affect your life?' (National Geographic, 2017, p.6). Iantaffi and Barker prepare the reader for what may prove a frightening descent into the unconscious:

> Take a moment to breathe and to think back about your early years, as far back as you can go. Don't worry if it's not too far at all – that's OK. Can you remember one of the first times that you realized what your gender was? It's OK if you can't. It can be as simple as being divided into girls and boys in school to line up, or for an activity, *and having an inner sense of where you belonged*. That inner sense may or may not have matched what others expected of you. Take a breath and notice what sensations, thoughts, and emotions emerge as you let yourself spend some time with those memories. (Iantaffi & Barker, 2017, p.54; emphasis added)

In a Foreword to the *Gender Quest Workbook*, transgender educator Ryan K. Sallans offers another primal gender memory:

> My first memory of gender, and what felt comfortable for me, took place when I was around three years old. It happened while I was standing outside, bare feet on the hot cement, next to a pool. (Testa, Coolhart & Peta, 2015, p.vii)

According to the reconstituted memory, Ryan discovers that he 'wasn't like the other boys' because '[m]y body was a girl's body'. The word 'comfortable' carries with it the promise of a return to childhood freedom: some trans men hope to regain a pre-pubertal bodily freedom after mastectomy.

Gender identity workbooks take the idea of a primal scene from Freud but recast it as a memory of gender. In their *Studies on Hysteria*, Breuer and Freud explain that 'in girls anxiety was a consequence of the horror by which a virginal mind is overcome when it is faced for the first time with the world of sexuality' (1974, p.192). Breuer and Freud believe that 'impressions from the pre-sexual period which produced no impression on the child attain traumatic power at a later date as memories' (p.200). Katharina's anxiety attacks turn out to have been triggered by the sight of her uncle 'lying on' her cousin, Franziska (p.193). Katharina goes on to describe a memory of her uncle making a sexual advance to her when she was fourteen. Where trauma is seen in terms of sex, the gender identity workbook can only see gender. Sallans describes 'growing crushes on girls', but *The Gender Quest Workbook* is not open to the possibility that this is an account of an emergent lesbian identity. Although the gender identity workbook promises the teenage reader 'your private space to explore all of you, inside and out' (Testa, Coolhart & Peta, 2015, p.vii), the space it provides is neither private nor free. Self-exploration takes place according to a preordained path which does not allow adequate space for the understanding of emerging sexual attraction. Nor does it encourage the recognition of the role of trauma or the 'cross-generational transfer of trauma' where a parent's early experience of bereavement may shape their child's self-understanding (Coates & Moore, 1998, p.39). In the brave new world of the gender identity workbook, neither sex nor death exist.

Identity in a bookstore

As a product of queer studies, the gender identity workbook privileges language as the place of identity construction and identity rebirth. A performative, as Judith Butler explained, is 'that discursive practice that enacts or produces that which it names' (1993, 2011, p.xxi). This is not really a difficult concept: as Butler points out: 'Let there be light: and there was light', the divine command with which Genesis opens, is a perfect example of performative language. Words, in this case, make something happen, and in the world of self-identified gender conjured by the gender identity workbook every individual can create their own identity through language alone. But Butler also makes it clear, following Derrida, that performative statements depend for their authority on pre-existing statements: only because we are citing or quoting earlier statements do our listeners recognize that we are making something happen. Freud's Katharina only comes to know herself through Freud's formulation, and Ryan K. Sallans discovers his identity from 'a book about transgender men in a bookstore'. That 'profound, inexplicable, intrinsic self-knowing' described by Serano depends on a conceptual framework transmitted through books.

The need to provide language explains the curiously adult technical language used by the workbooks. The three to five-year-old addressed by *Who are You?* learns 'just a few words people use: trans, genderqueer, non-binary, gender fluid, transgender, gender neutral, agender, neutrois, bigender, third gender, two-spirit' (Pessin-Whedbee, 2017, p.20). (Has the author met a three-year-old, I wonder?) The seven-year-old reader of *The Gender Identity Workbook for Kids* is primed to pass the key diagnostic test of a transgender identity by Activity 16 which introduces 'The Three 'Tents': ConsisTENT, PersisTENT, InsisTENT' (Storck, 2017, p.66).

These three words are used by US affirmative gender therapists to distinguish 'transgender' from 'gender diverse' children: 'Transgender and gender diverse children may exhibit similar preferences', explains a Fact Sheet produced by the American Psychological Association. 'A pervasive, *consistent, persistent and insistent* sense of being the other gender and some degree of gender dysphoria are unique characteristics of transgender children' (Meier & Harris, n.d.; emphasis added). As the workbook explains: 'Those are some big words! Let's look at what they mean'. The gender identity workbook does not understand speech act theory or recognize the necessarily iterative nature of performative utterances but offers to identify a unique self through technical language which prepares the child for social transition and may lead to pharmaceutical and surgical modifications of the body.

The Gender Obsession

Whether addressed to a three-year-old or an adult, the first project of the workbook is to convince the reader of the supreme importance of 'gender', a task achieved by repetition of the talismanic word. *You and Your Gender Identity: A guide to discovery* by Dara Hoffman-Fox (2016) uses 'gender' 202 times in a book of 280 pages; in its 284 pages, *How to Understand Your Gender* (Iantaffi & Barker, 2017) repeats the word 239 times. *The Gender Quest Workbook* for teens and young adults begins by asking: 'What is gender anyway?', 'Has gender always been around?', 'Will it always be around?', 'If I move to Jupiter, will gender exist there?' and 'Do all animals have genders?' (Testa, Coolhart & Peta, 2015, p.5). The seven-year-old picking up *The Gender Identity Workbook for Kids* encounters 'Section 1: Understanding Gender':

> Gender can be challenging to wrap your head around because many of us don't have all the information about what gender really is (or isn't), or we have information which isn't completely correct. (Storck, 2018, p.1)

Activity 1 teaches the seven-year-old to distinguish between 'assigned sex', 'gender expression' and 'gender'. The child's knowledge of their gender is supposedly preverbal, it is a 'deep-down feeling':

> Gender is your deep-down feeling that you're a boy, a girl, neither, both, or something else. That's right! Not just a boy OR a girl – some people know themselves to be BOTH boy and girl or NEITHER boy nor girl or SOMETHING ELSE altogether. There are so many genders, and the expert on your gender is YOU! (Storck, 2018, p.3)

Yet this knowledge must be conveyed in terms provided by the workbook. Capitalization and exclamation marks reveal the authoritarian tone, deployed paradoxically to insist that the child is the expert: 'You are the expert when it comes to your gender!' (Storck, 2018, p.2), 'the expert on your gender is YOU!' (Storck, 2018, p.3). The section ends with a revision exercise with prescribed answers, a modern-day catechism:

> **For You to Do**
> These are really important ideas, so let's check that we are on the same page.
> 1. How do people assign a baby's sex?
> 2. What do we call the way a person *expresses* their gender?
> 3. Who is the absolute best person to know and declare your gender?
>
> **Answers**
> 1. By looking at the baby's body. Remember, this assigned sex is not always the same as a person's gender!
> 2. Gender expression.
> 3. YOU!!!

At seven, kids may not understand the biology of reproduction and are therefore especially vulnerable to misinformation. They have no means of rejecting the claim that '[p]eople have a mistaken belief that having a penis means you are a boy and having a vulva means you are a girl', a claim that is central to all the workbooks, for it depends on a central tenet of the gender revolution: that sex (or rather the sexed body) is a cultural construction. Writing for adults, Iantaffi and Barker challenge knowledge of biological reality by the repeated assertion that sex is complex: 'We know that we're using the word 'complex' a lot here', they say, 'but it really reflects reality!' (Iantaffi & Barker, 2017, p.46). It is 'very difficult to know what sex we really are' (Iantaffi & Barker, 2017, p.40). Not only are these claims counterintuitive, they are also factually wrong. London's Gender Identity Development Service (GIDS), treating the largest population of children with gender dysphoria in the world, has found that gender dysphoria does not correlate with intersex conditions or disorders of sex development: 'An audit of UK clinics from 2013 to 2015 [...] revealed no differences from cytogenetic surveys of the UK newborn population and elsewhere. Therefore, routine karyotyping [chromosomal analysis] of a child or adolescent with GD is not required unless any specific clinical features determine this to be necessary' (Butler et al, 2018). The idea of 'complexity' enables the authors to assert academic authority in the face of a reader's intuitive scepticism: Iantaffi and Barker confidently assert that 'penises and vaginas are not inherently male or female' (2017, p.40). Biological sex has been swept away by the new (but supposedly age-old) knowledge of the gender revolution.

Sex as cultural construction

The claim that sex is a cultural construction derives from two books published in 1990: Judith Butler's *Gender Trouble* and Thomas Laqueur's *Making Sex: Body and Gender from the Greeks to Freud*. Butler and Laqueur, in their turn, derive their ideas from the cultural historian Michel Foucault's *The History of Sexuality* (1976, tr. 1979). In addition, Butler draws on ideas from Mary Douglas to arrive at the idea that we cannot access an idea of the body outside of culturally constructed taboos (Butler, 1990, 2006, p.179). These ideas filter into transgender studies: in a 2007 article, Stephen Whittle and Lewis Turner cite Laqueur to claim that 'originally the categories 'male' and 'female' were understood as residing in one body and sex was a sociological category rather than a biological one'. Although unacknowledged, Laqueur is also the source for Iantaffi and Barker's remark that 'people are often surprised to learn that for some time there was assumed to be just one gender, with women being a slightly inferior kind of man' (2017, p.68). Whittle and Turner argue that the 2004 Gender Recognition Act reversed the assumed priority of sex over gender so that 'Gender [...] now determines 'sex'', reinstating a hierarchy of categories which existed prior to the nineteenth century. Just as the sexual revolution of the twentieth century allowed the return of old sexual freedoms, so the argument goes, the gender revolution of the twenty-first century reinstates older wisdom about the relationship of sex and gender.

But Laqueur's argument was misunderstood by Whittle and Turner and by the transgender history that developed in their wake. It is true that Laqueur shows how culture shapes our understanding of the body: he argues that anatomy is 'not pure fact, unadulterated by thought or convention, but rather a richly complicated construction based not only on observation, and on a variety of social and cultural constraints on the practice of science, but on an aesthetics of representation as well' (1990, pp.163-4). But Laqueur also insists that the realm of 'fact' exists, that 'an anatomy text or illustration' can be judged 'more or less accurate' and that there is 'progress in anatomy' (1990, p.164). Laqueur's point is that culture always inflects the way in which we understand biology. Thus the older 'one sex' model (under which the female genitals were seen as an inverted image of the male, and women were a wetter, weaker and colder version of the male body) was 'framed in antiquity to valorize the extraordinary cultural assertion of patriarchy, of the father, in the face of the more sensorily evident claim of the mother' (1990, p.20). In appealing to the 'one sex' model to support a paradigm shift, Whittle and Turner are attempting to reinstate a patriarchal model.

According to Laqueur, cultural imperatives also shape our understanding of the natural world. With a growing understanding of the mechanics of sexual reproduction at the end of the eighteenth century, he explains, 'sex also filtered down from animals to plants' (1990, p.172), inflecting a new understanding of botany according to Linnaean sexual taxonomies. The seven-year-old reader of *The Gender Identity Workbook for Kids* is required to reverse this process to misunderstand human reproduction by analogy with frogs, lizards and butterflies. The child is told that human beings reproduce like seahorses: 'Unlike most animals, the male seahorses become pregnant and give birth to baby seahorses. This is also true for humans. Having babies isn't just for moms – transgender dads sometimes give birth to babies also!' (Storck, 2018, p.13). The sexual revolution of the twentieth century enabled children to access human biology rather than euphemistic accounts of the loves of the birds and the bees. But the gender identity workbook denies the child reader an understanding of mammalian reproduction familiar to children at the beginning of the twentieth century. Freud recalls a small boy who 'said to his little sister, 'How can you think that a stork brings babies! You know that man is a mammal, do you suppose that storks bring other mammals their young too?'' (1963, p.23). Far from taking Laqueur's point that knowledge is culturally constructed, the workbooks of the gender revolution use the natural world to buttress counterfactual claims about biology. If Laqueur were to read the workbooks, my bet is that he would see them as performing the work of cultural construction he recognized in the older medical tradition. Just because these books are new does not mean they escape cultural construction.

Pick a gender, any gender

The practical exercises prescribed by the workbooks involve trying out a series of labels: the reader is asked how 'would it or does it feel when people see you as a gender other than girl/woman or boy/man (for example, as androgynous or Two-Spirit)?' (Testa, Coolhart, & Peta, 2015, p.21). Iantaffi and Barker provide a box containing 'all the words that Facebook was offering for gender when we were writing this book' with the injunction to 'circle any words that feel like a good fit for you' (2017, p.36). This process recalls the superstitious reading practices of earlier centuries by which the uneducated would open a copy of Virgil and interpret their fate according to the first word they saw (Pearson, 1999, p.94). The child reader is also invited to contemplate the example set by model children. *The Gender Identity Workbook for Kids* describes agender Jaquise who 'grew up with everyone calling them a girl' but 'knew this didn't feel right'; Alex, who is 'a proud and creative person' who 'doesn't feel all boy or all girl' and calls themselves 'gender-fluid' or 'genderqueer"; and Micah, who 'loves all things bright, sparkly, fancy and fabulous' and is a 'gender-expansive Kid who sometimes calls himself a Princess Boy' (Storck, 2018, p.10). These ideal figures may demonstrate a new set of gender virtues but in other ways resemble nothing more than the children of earlier religious literature: Sarah, from James Janeway's *A Token for Children*, for instance, 'was very conscientious in spending of time, and hated idleness, and spent her whole time either in praying, reading, instructing at her needle, at which she was very ingenious' (Janeway, 1793, p.23). Like the stories of the saints on which earlier generations were raised, the stories of gender journeys demand an intense process of self-examination during which the self is reshaped according to ideal models. But self-actualization is achieved via medical technologies which were not available to Joan of Arc, cited anachronistically as transgender in the *Gender Identity Workbook for Kids* (Storck, 2018, p.19).

Just as history is ransacked for examples of gender nonconformity which can be reconstructed through the twenty-first century lens of transgender identity, the first world child reader is encouraged to pick and choose selves from the enticing array of non-western cultures, stripped of cultural context. The gender revolution was announced in *National Geographic*, a magazine that speaks from an established position to a global audience, which has been published continuously since 1888 and is now available in 37 local languages around the world. The gender identity workbook, together with popular journalism, recycles the work of trans writers such as Leslie Feinberg, whose 1996 *Transgender Warriors* describes combing through 'books, periodicals, news clippings devoted to the history of Europe, Africa, Latin America, the Middle East, and Asia' in a search 'for the earliest written records of any forms of trans expression' (1996, p.40). But in doing so, the gender identity workbook misunderstands this search. Feinberg movingly describes how the discovery of diverse identities freed her of 'a layer of unexamined shame' associated with her own same-sex desire: it is clear that Feinberg is writing about sexuality (1996, p.39). Feinberg does not employ the rhetoric of gender identity but describes a pragmatic decision: 'My life changed dramatically the moment I began working as a man', Feinberg explains: 'I was free of the day-in, day-out harassment that had pursued me' (1996, p.12). A neo-liberal revision of this powerful narrative strips away the context of sexual desire and serves up Feinberg's transgressive knowledge as facts to be learned by rote by children: 'Activity 5: Gender around the world' in *The Gender Identity Workbook for Kids* represents complex social practices as choices (Storck, 2018, p.15). The gender journey mapped by the workbooks and by Katie Couric's documentary is a new colonial journey in which the technological advances of twenty-first century medicine are disseminated via didactic literature to a world in thrall to American culture. The metaphor recalls the colonial exploration on which this culture is based.

In search of the authentic self

One origin of the journey to find 'your unique gender' (Storck, 2018, p.1), may be found in the American west coast search for 'a new form of the individual and a new form of individualism' born in the 1960s (Storr, 2017, p.128). The Esalen Institute taught that 'God lay in the deep, authentic core of everyone', that the true self within must be actualized (Storr, 2017, p.121). At Esalen, adults confronted a radical authenticity through re-experiencing their own births. The first entry for 'transgender' in the MLA international bibliography is Susan Stryker's 1994 essay 'My Words to Victor Frankenstein above the Village of Chamounix: Performing Transgender Rage', a work which announces both the birth of the new academic discipline of transgender studies and the birth of a new gendered self: 'Battered by heavy emotions, a little dazed, I felt the inner walls that protect me dissolve to leave me vulnerable to all that could harm me. I cried and abandoned myself to abject despair over what gender had done to me' (Stryker, 1994, p.248). 'In birthing my rage, my rage has rebirthed me'.

Whereas Stryker tells a story of liberation through rage, Serano recalls an attempt to suppress intrusive thoughts. The result was predictable: 'The more I tried to ignore the thoughts of being female, the more persistently they pushed their way back into the forefront of my mind' (Serano, 2016, p.82). As in Winerman's famous 'white bear' experiment into thought suppression, the thought returned with added force ('unwanted thoughts'; 2011). But transition did not cure the obsession. Serano writes that:

> I still struggle with an intense hypersensitivity to gender (and more specifically to gendering). [...] And although I experience gender concordance these days, I still constantly dwell on gender, which, while

helpful when writing a book on the subject, can often be unhealthy and exhausting. (2007, p.182)

Rather than countering this 'unhealthy and exhausting' obsession, the gender identity workbooks encourage rumination. As McGee argues, 'paradoxically [self-help] literature may foster, rather than quell, their anxieties' (McGee, 2005, p.17). The seven-year-old who was introduced to the term 'gender' in Activity 1 ends by writing a letter to the malign deity that the book has created—Activity 36, 'More to Do', tells the reader: 'Sometimes it can be hard to quiet the pressures of your gender. If that is true for you, try asking your gender to take a rest. Go ahead, write a note to let your gender know you need some time to enjoy or deal with something else' (Storck, 2018, p.142). A young detransitioner describes the compulsion that came with her trans identity:

> They told us that we could choose a gender, any gender, out of countless, that we could make up our own and they would be taken seriously; they were, but only ever by others on there. Words on Tumblr ceased to mean the same as in the real world. Words were made up. [...] If we liked to switch how we 'presented', we would have a label to describe that we switched, and we could also change our labels and our pronouns day-to-day to describe how we felt (FELT! That is the crux of all of this nonsense) each day. It is so, so exhausting to be constantly examining every desire, thought, inclination of your shifting, constantly changing adolescent self, trying to find a word to fit, only to question yourself again the next week, or day, or hour. (Sam, 2018)

The search for the authentic self born on the west coast of the US does not derive from Foucault, Laqueur or Butler. Foucault attacked the language of 'internalization'—the idea that our true self is inside us, 'what he took to be the psychoanalytic belief in the 'inner' truth of sex' (Foucault, 1976, p.183). As Butler wrote in 1990: 'If the inner truth of gender is a fabrication and if a true gender is a fantasy instituted and inscribed on the surface of bodies, then it seems that genders can be neither true nor false, but are only produced as the truth effects of a discourse of primary and stable identity' (Butler, 1990, p.186). Butler talks about the 'illusion of an abiding gendered self' (1990, p.191); the title of her 2004 book, after all, is *Undoing Gender*.

Foucault's history of gender

It is not surprising, then, that Foucault provides the best account of the gender ideology disseminated by the gender identity primers and workbooks of the twenty-first century. The story of the gender revolution told by the *National Geographic* and the gender identity workbooks echoes the historical narrative that underpinned the sexual revolution of the twentieth century. As Foucault explained in 1976, the sexual revolution offered to rediscover freedoms lost under capitalism: 'By placing the advent of the age of repression in the seventeenth century, after hundreds of years of open spaces and free expression, one adjusts it to coincide with the development of capitalism: it becomes an integral part of the bourgeois order' (1976, p.5). Bourgeois society represses sexual freedom, according to this story, and sexual transgression is an attack on bourgeois morality. Foucault noticed the 'solemnity with which one speaks of sex nowadays', with the prophetic tone:

> ...we are conscious of defying power, our tone of voice shows that we know we are being subversive, and we ardently conjure away the present and appeal to the future, whose day will be hastened by the contribution we believe we are making. Something that smacks of revolt, of promised freedom, of the coming age of a different law, slips easily into this discourse on sexual oppression. Some of the ancient functions of prophecy are reactivated therein. Tomorrow sex will be good again. (1976, pp.6-7)

We hear this prophetic tone in the prophets of the gender revolution: tomorrow, today even, gender will be good again.

But Foucault did not subscribe to the story of escape from repression. He pointed out that throughout the period of supposed repression, people talked about sex because they had to. They were told to confess every sexual thought and action (and specify the details). Sleeping arrangements were monitored, sexual practices reviewed. Autobiographies revealed details of sex lives. Doctors, moralists, churchmen and psychiatrists created complex taxonomies of sexual practices. '[A]round and apropos of sex', Foucault says, 'one sees a veritable explosion' (1976, p.17). Sex was 'endowed [...] with an inexhaustible and polymorphous causal power' (1976, p.65). According to Foucault, the discourse of sexual liberation offered only a continuation of this process, once again shaping and directing the expression of sexuality, forcing it to conform to a new narrative. Sexologists like Krafft-Ebbing, Tardieu, Molle, and Havelock Ellis 'carefully assembled this whole pitiful, lyrical outpouring from the sexual mosaic' (1976, p.64). Instead of sexual self-examination, the twenty-first century reader is encouraged to believe that we need 'to do reflective work' on our gender. This work is a new form of the spiritual diary, the daily stock-taking in which the individual counted sins and named sexual faults. Sex used to be the private place where we became our innermost self. Gender now provides the unique code that makes us into an individual:

> Given that everybody's biological make-up, psychological experiences, and social context connect up in unique and complex ways, our gender really is something like a snowflake: no two of us are quite the same. (Iantaffi and Barker, 2017, p.46)

Today it is the gender specialist who speaks of a new 'Gendered Intelligence'. These are Foucault's 'perpetual spirals of power and pleasure'.

There is a lot about Foucault's writing that is difficult to test empirically. But we can see the compulsion to talk about sex, the 'veritable explosion' he describes if we look at the frequency of the word 'sex' and 'sexuality' across different discourses. As the nineteenth century gave way to the twentieth, a compulsion to talk about sex spread from religion to medicine, to sexology and to many other academic disciplines. In 2000 Bruce R. Smith did a search for the word 'sexuality' from 1981 onwards on work listed on the online bibliography of the Modern Language Association (MLA), an organization dedicated to the study and teaching of language and literature. He found 'well over three thousand items' (Smith, 2000, p.318). When we repeated the experiment in 2018, we found that the words 'sex' and 'sexuality' had been overtaken by 'gender': whereas the MLA international bibliography now gave us 3,230 results for 'sexuality' in titles since 1981 (with 4,039 results for 'sex' since 1981), a comparable search for 'gender' produced 14,355 results. 'Gender' has taken over as the key explanatory term of the twenty-first century, outpacing 'sex' and 'sexuality' by about three to one, beginning its rise from 1990 onwards to dominate articles written in the twenty-first century. Other keywords such as 'queer', associated with Judith Butler's 1990s work, are now challenging the predominance of 'gender', with 'queering' part of the academic arsenal since 1994.

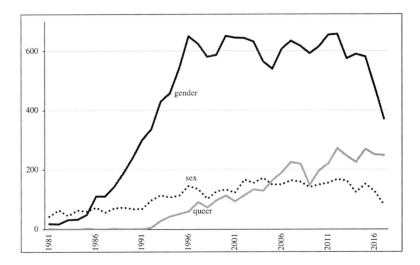

Figure 14-1: 'Gender', 'Sex' and 'Queer' keyword search on the MLA international bibliography

If we follow Foucault's logic, it is not the reawakening of a buried gender freedom that we see in the explosion of gender texts but the production of discursively shaped gender identities. You only have to replace 'sex' with 'gender' to generate from Foucault's history a perfect account of the disciplinary function of the new religion of gender: '[A]round and apropos of [gender]', Foucault might say if he were alive now, 'one sees a veritable explosion'. For gender is now 'endowed [...] with an inexhaustible and polymorphous causal power' (1976, p.65). What we are seeing is the creation of a new system of knowledge, a system that is a form of discipline:

> This society that emerged [...] did not confront [gender] with a fundamental refusal of recognition. On the contrary, it put into operation an entire machinery for producing true discourses concerning it. Not only did it speak of [gender] and compel everyone to do so; it also set out to formulate the uniform truth of [gender]. As if it suspected [gender] of harboring a fundamental secret. As if it needed this production of truth. As if it was essential that [gender] be inscribed not only in an economy of pleasure but in an ordered system of knowledge. (Foucault, 1976, p.69)

Iantaffi and Barker (and their like) are not heirs of Foucault or Butler, but their misreaders.

Acknowledgements

I would like to thank Michael Biggs who prepared the graph and provided helpful advice on an earlier draft of this chapter.

References

Bornstein, K. (1998). *My Gender Workbook: How to Become a Real Man, a Real Woman, the Real You, or Something Else Entirely.* New York: Routledge.
Bornstein, K. (2013). *New Gender Workbook: A Step-by-Step Guide to Achieving World Peace through Gender Anarchy and Sex Positivity.* New York and Abingdon: Routledge.
Breuer, J. & Freud, S. (1974). *Studies on Hysteria,* tr. James and Alix Strachey. Harmondsworth: Penguin.
Burns, C. (Ed.). (2018). *Trans Britain: Our Journey from the Shadows.* London: Unbound.
Butler, G., De Graaf, N., Wren, B., and Carmichael, P. (2018). Assessment and support of children and adolescents with gender dysphoria. *Archives of Disease in Childhood, 103,* 631-636.
Butler, J. (2006) [1990]. *Gender Trouble.* New York and London: Routledge.
—. (2011) [1993]. *Bodies that Matter: On the discursive limits of 'sex'.* London and New York: Routledge.
Coates, S.W. & Moore, M.S. (1998). The complexity of early trauma: representation and transformation. In D. Di Ceglie and D. Freeman (Eds.), *A Stranger in my Own Body: Atypical Gender Identity Development and Mental Health* (pp.39-62). London: Karnac.
Feinberg, L. (1996). *Trans gender Warriors: Making History from Joan of Arc to Dennis Rodman.* Boston: Beacon Press.
Foucault, M. (1979) [1976]. *The Will To Knowledge, The History Of Sexuality Part I.* Tr. Robert Hurley. London: Penguin.
Goldman, M. (2012). *The American Soul Rush: Esalen and the Rise of Spiritual Privilege.* New York: New York University Press.
Hausman, B.L. (1995). *Changing Sex: Transsexualism, Technology, and the Idea of Gender.* Durham, NC: Duke University Press.
Hines, S. (2018). *Is Gender Fluid? A Primer for the 21st Century.* London: Thames and Hudson.
Hoffman-Fox, D. (2016). *You and Your Gender Identity: A guide to discovery.* Colorado Springs: DHF Press.

Iantaffi, A. & Barker, M-J. (2017). *How to Understand Your Gender: A practical guide for exploring who you are*. London: Jessica Kingsley.

Janeway, J. (1793). *A token for children: Being an exact account of the conversion, holy and exemplary lives, and joyful deaths of several young children*. Glasgow.

Laqueur, T. (1990). *Making Sex: Body and Gender from the Greeks to Freud*. Cambridge, MA, and London: Harvard University Press.

McGee, M. (2005). *Self-Help, Inc.: Makeover Culture in American Life*. Oxford: Oxford University Press.

Pearson, J. (1999). *Women's Reading in Britain 1750-1835*. Cambridge: Cambridge University Press.

Pessin-Whedbee, B. (2017). *Who are You? The kid's guide to gender identity*. London and Philadelphia: Jessica Kingsley.

Serano, J. (2016) [2007]. *Whipping Girl: a transsexual woman on sexism and the scapegoating of femininity*. Berkeley, CA: Seal Press.

Smith, B. R. (2000). Premodern Sexualities. *PMLA, 115*(3), 318-29.

Storck, K. (2018). *The Gender Identity Workbook for Kids: A Guide to Exploring Who You Are*. Oakland, CA: New Harbinger.

Storr, W. (2017). *Selfie: How we Became so Self Obsessed and What it is Doing to Us*. London: Picador.

Stryker, S. (1994). My Words to Victor Frankenstein above the Village of Chamounix: Performing Transgender Rage. *GLQ: A Journal of Lesbian and Gay Studies, 1*, 237-254.

Testa, R.J., Coolhart, D. & Peta, J. (2015). *The Gender Quest Workbook: a guide for teens & young adults exploring gender identity*. Oakland, CA: New Harbinger.

Whittle S. & Turner, L. (2007). 'Sex Changes'? Paradigm Shifts in 'Sex' and 'Gender' Following the Gender Recognition Act? *Sociological Research Online, 12*(1). doi:10.5153/sro.1511

Online Sources

Katie Couric Media. (2016). *Gender Revolution: A Journey with Katie Couric* [video file]. National Geographic Channels. Retrieved from https://www.youtube.com/watch?v=7u3YO2CJNcg

Meier, C. & Harris, J. (n.d.). Fact Sheet: Gender Diversity and Transgender Identity in Children. *American Psychological Association*. Retrieved from https://www.apadivisions.org/division-44/resources/advocacy/transgender-children.pdf

National Geographic (2017). *Gender Revolution. A Discussion Guide for Teachers and Parents*. Retrieved from

https://www.nationalgeographic.com/pdf/gender-revolution-guide.pdf
Sam (2018, March 12). Baptised in Fire: A relieved desister's story. *4thWaveNow*. Retrieved from https://4thwavenow.com/
Steinmetz, K. (2014, May 29). The Transgender Tipping Point. *TIME*. Retrieved from http://time.com/
Uglow, T. (2018, March 8). We understand the solar system, so why do people still struggle with gender? *The Guardian*. Retrieved from https://www.theguardian.com/
Winerman, L. (2011). Suppressing the 'white bears'. *Monitor on Psychology, 42*(9). Retrieved from https://www.apa.org/monitor/2011/10/unwanted-thoughts.aspx
'workbook, n.'. *OED Online*. (J2018). Retrieved from http://www.oed.com/view/Entry/416183

CHAPTER FIFTEEN

RAPID ONSET GENDER DYSPHORIA

MICHELE MOORE

Rapid Onset Gender Dysphoria (ROGD) is the sudden declaration of transgender identity by children and young people. The purpose of this chapter is to pose questions about the growing phenomenon of ROGD and to explore the ways in which, and reasons why, transgender activists attempt to close down research about it. The silencing of debate is more than discursive in that it threatens not only our independence as researchers, but also materially enters into the arena of family life and clinical intervention.

The chapter engages with three sources of evidence for ROGD:

- the ground-breaking work of Lisa Littman which has established that ROGD is *not* an indicator of transgenderism (Littman, 2018);
- new data drawn from a series of open-ended accounts written by parents of trans-identified children and young people involved with the parent-led organization Gender Identity Challenge Scandinavia (GENID); and
- the views of clinicians from gender identity development services in the UK and Scandinavia recorded by myself and my co-editor Heather Brunskell-Evans.

The analysis arising from the juxtaposition of these three sources indicates remarkable agreement between the observations of researchers and parents and the concerns of gender-critical clinicians about the perilous consequences of denying the actuality of how ROGD invents the transgender child.

Rapid Onset Gender Dysphoria is not an indicator of transgenderism

Littman's evidence indicates that gender dysphoria can represent a maladaptive coping mechanism for distress or trauma and that peer and social influences, including online induction and coaching, are common pre-determinants of transgender feelings. These factors may interact in producing and maintaining gender confusion. Parents of transgender-identifying teenagers told Littman that if they did not immediately affirm their child as transgender they were quickly marginalized and ultimately excluded from having any input into intervention planning and care by their own children, clinicians and other professionals including teachers and social workers. These findings are of vital importance because an amalgamation of exposure to online influences, combined with the barring of the knowledge of parents, frames the practice of clinical intervention in ways which intersect with the aspiration of transgender activists that they should be the gatekeepers of transgender diagnosis and intervention.

The suppression of evidence that builds understanding of ROGD is therefore fundamental to the maintenance of transgender ideology that transgenderism is inherent and that children whose parents do not automatically affirm must be protected from their parents. The actuality of ROGD is, conversely, fundamental to understanding the incalculable dangers inherent in the invention and promotion of 'the transgender child'.

Suppression of research on Rapid Onset Gender Dysphoria

I collaborated closely on the development of the research, analysis and minutiae of Littman's seminal 2018 paper. Her empirical findings drew attention to the growing phenomenon of rapidly occurring transgender identification observed by parents, which had not hitherto been studied systematically in the scientific literature. What happened to Littman's publication of these findings tells a story about how transgender activists seek to disrupt independent research and how this currently functions not only to compromise the integrity of research and its institutions but also to block clinical excellence for children and young people struggling with gender dysphoria. However, the process of bringing Littman's data to light has not entirely been one of negativity and setback. It has spearheaded a push-back from the community of researchers and parents trans activists wish to regulate, which is taking us beyond the determinism of their agenda. Documenting the ways in which research about ROGD is being suppressed enables recognition of its explanatory power and significance for the domain of transgender intervention, and of the importance of a collective refusal to collude in rendering the accounts of parents and young people as either invisible or pathological.

Following publication, Littman's paper was immediately denounced by critics including reviewers who said they had not even read it. The paper received a tsunami of predictably reductive responses from trans activists intent on closing down legitimate discussion by attacking researchers rather than engaging with the substance of the research itself. Littman was accused of employing partisan research methods to promote harmful misconceptions about transgender people.

In response to complainants, the publishing journal *PlosOne* initiated a post-publication re-review of Littman's paper, thereby, extraordinarily, suggesting to the scientific community that the integrity of its in-house staff editors, international board of academic editors and the independent peer review process could be called into question by ideological objectors; Littman had already recognized and articulated the limitations of her work in the published paper, in accordance with the highest standards of critical scholarship exacted through several rounds with PlosOne reviewers. Perhaps the editors were intimidated by the complainants and instead of relying on their own judgment, felt compelled to add more and more external reviewers to guide them. Perhaps the editors were unaware that some of the comments and complaints were not made in good faith.

Littman's own Ivy League university, contradicting its public mission to 'serve the community, the nation, and the world by discovering, communicating, and preserving knowledge and understanding *in a spirit of free inquiry*' (my italics) took the paper down from its own publication sites. Neither *PlosOne* Editor-in-Chief Joerg Heber nor Bess Marcus, Dean of Brown University's School of Public Health, acknowledged the implications of possible collusion with activists attempting to close down free academic speech (personal communications). Both institutions were arguably complicit in rendering the social and psychological context of children and young people's identity impervious to reasoned reflection. Littman was asked to turn in numerous additional rounds of review and to address criteria that had not previously been a prerequisite for publication of qualitative and mixed-methods studies in *PlosOne*. Months later however, following the conclusion of the comprehensive *PlosOne* review, an expanded version of her paper was published making clear there can be no doubt that the study is reported in sufficient detail to enable a clear understanding of how the data were treated and interpreted (Littman, 2019). The findings of Littman's paper have thus been substantially vindicated by extensive, repeated and rigorous review: there is no question that her research and its conclusions are robust. Her headline finding, that ROGD may invalidate a clinical 'transgender' diagnosis for a child or young person, stands.

One of Littman's hoped-for outcomes for the publication was to open up rather than foreclose public debate. By drawing attention to the push to close down debate through ad hominem charges, the paper exceeded the scope of its original intentions. Hundreds of thousands of views and downloads within months of the paper appearing verify the interest, reach and significance of Littman's work for the scientific community and other research users.

Parents challenging gender identity in Scandinavia

Denial of ROGD means the opportunity to address the origins and significance of gender distress is not only missed but also locks a child or young person into a mistaken invention of themselves as 'transgender'. Stories shared by the GENID parents from four countries in Scandinavia offer not only additional data with which to expand Littman's findings and analysis, but also perspectives from a previously under-researched social and cultural context (Brunskell-Evans, 2018). The parents asked to share their own experience to make visible the reality of ROGD in their own families, and to testify to devastating outcomes of intervention based on self-affirmation of transgenderism which denies ROGD, that they are witnessing first-hand. Their stories extend Littman's preliminary findings to show the consequences of ignoring ROGD in favour of promoting 'gender identity' as a 'truth' to be deployed in a clinical setting are potentially devastating, life-long and life-ending.

The perspectives of parents are also connected to the concerns of anonymous clinicians from specialist services. It may seem surprising that clinicians who are so experienced with issues of trauma and social contagion underlying self-expression and self-determination of distress in children and young people say they are not addressing these issues for those presenting gender dysphoria. They explain this by saying they come under pressure from lobby groups not to use their own training and to cast aside the requirement for research-led clinical thinking about the developing brain:

> The lobbyists oppose, for example, a neuro psychological model of child development which we all learned as clinical psychologists, foundational starting points such as Piaget's work on the child's ability to think abstractly, knowledge about the adolescent brain, about risk taking ... all

of that 'truth' we use in all of our work with children and young people we are inexplicably *not* using in this service. (Anonymous gender clinician, 2018)

What are the implications of specialist gender clinicians being unable to use their clinical training in practice with children who identify themselves as transgender? What are the implications of specialist gender clinicians, knowing that children and young people referred for gender dysphoria may have a substantial co-occurring history of psychosocial and psychological vulnerability, being unable to carry out a comprehensive assessment that goes beyond immediate affirmation of transgenderism?

The material assembled for this chapter indicates agreement between observations of parents and concerns of gender-critical clinicians stemming from different vantage points and signifies a multiplicity of concerns about the consequences of not exploring ROGD. Parent contributors catalogue a series of devastating outcomes for their children who have been diagnosed as transgender, including, for one family, post-transition death by suicide. The parents' accounts reveal appalling dilemmas about the trust they can employ in gender clinicians who insist a self-affirmation of transgenderism is adequate for clinical diagnosis and actively turn their intelligence away from opposing evidence. Their experience reveals a travesty of expertise vital for the proper support of gender-confused children and young people. They insist that the inclusion of parents' perspectives in the diagnosis of children is essential to balance over-reliance on self-identification. This is not to say parents should be exclusively responsible for confirming or refuting their child's self-identification as transgender, but that the current practice of excluding parents from the clinical picture collaborates in the invention of the transgender child and in actively promoting and reinforcing transgenderism.

Contesting the origins of transgender identification

The link between social media and 'out of the blue' announcements of being transgender has been one of the most powerful findings reported by Littman for understanding gender identity issues. What parents say about their child's transgender identification enriches the understanding of how ROGD has erupted across the landscape of children and young people's lives via social media:

Anna's parents:

> When she was eleven Anna got her own smartphone and stopped being sporty, stopped going out, stopped having friends on visits. When the smartphone came she spent more and more time in the room with the phone. She became in a way 'not present'. We don't know what happened but she went through a comprehensive change of behavior in a short period of time. She began to be interested in Japanese comics, Cosplay, made new online friends. Then soon after, one evening, she said 'Mom, I'm a Boy'.
>
> We did not know what to think. Was this just a phase? She is not a boy and she never showed any signs of being a boy. Then she sent us a text message:
>
> *Hello, there's something I've been thinking about telling you, I feel like a boy. Therefore, I want everyone to use 'he' as my pronoun and use the name Jaakko. This is very important to me, and I know it can be difficult for you who are cis people to understand, but I expect you to try, although it will certainly take some time to get used to it. Regards Jaakko.*
>
> What struck us was that this message was so unnatural. It did not sound like Anna.

These observations imply that instructions gleaned on the internet help construct the child's invention of themselves as transgender. Instructions accessed online are then used to regulate everyday acts so that the gender identity is reinforced (suggesting that otherwise it might not be stable):

> We received instructions and 'training' by Anna, in how to talk to people who had a different perception of gender than us. We accepted that she changed her hair and style. She talked to her school and they complied with Anna.

Anna's exposure to online influence is obvious:

> I logged in on her computer out of desperation to know what could be wrong. I learned that Anna has been part of a larger network of young girls who have been fascinated by popular YouTubers who say how nice it is to change sex and how wonderful it is to take testosterone and share advice on how to 'pass' as a transman. She was accessing a whole world of 24-hour online support, run by and for transgender people. They send out free breast binders. They sell 'packers', imitation male genitals of all sizes and colors for children from age three years. I discovered a dark and suggestive subculture of transgender activism in which children and young people who are unsure of themselves are being groomed in trans-ideology. Anna, in a very vulnerable situation and age, had been groomed on social media, to become a boy.

Many gender clinicians are aware of the role of social media in the invention of the transgender child:

> ...the story of self-identification of transgenderism a young person is telling you will know, or suspect, is steeped in the influence of online social platforms like Twitter and Tumblr and Reddit. Every time. (Anonymous gender clinician, 2018)

Further, clinicians point to the influence of *'homophobia in the whole of culture, sexism and misogyny characteristic of social media platforms'*, which parents notice as part of the backdrop to their child thinking they are transgender:

Filip's mother:

> I asked Filip about what happened in the beginning, when he first started to think about his gender. He told me he wasn't satisfied with his body. He thought he was too thin. I also asked him about how he felt and thought about girls and the female gender. He told me that he thought girls looked nice with their more round and soft forms but that he mostly has been attracted to boys. Through the internet he found out about the possibility to change his body and became interested in this.

Mathias's parents:

> After he said he is transgender his younger brother searched for insight in his comment on the LGBT site. There are 383 comments and the three biggest categories are: having a bad body image, feeling ugly and fat and his desire for friends—real friends.

In the words of gender clinicians:

> ...yeah ... so basically, we hear all these stories of the internet and the influences that are out there and it's very hard to know whether you think there is any innate gender id. But the questions aren't allowed to be asked about that ... we can't ask certain things ... even going down this route of thinking is seen as being completely unacceptable, even to be thinking about it, questioning it... (Anonymous Gender Identity Services clinician)

> ...it's true, the lobby groups are creating an atmosphere of great difficulty ... we have to affirm, no questions asked ... we have to affirm, we have to affirm, we have to affirm... (Anonymous Gender Identity Services clinician)

There may, however, be no stability in a young person's self-identification as transgender, suggesting that a strategy of 'no questioning' may freeze the development of their thoughts and not offer support but consolidate confusion.

Mathias's parents:

> Sometimes he says he found out he was transgender at the age of 9, sometimes he tells us it happened in his early teenage years, he told the school it happened in the high school age 16-19. When he told me he was 17. He seemed nervous, downhearted and in pain when he told me. He had cut his arms. He sent an SMS saying 'I am transgender. I knew since the age of 9 that I was a woman in the wrong body' but we never saw any signs of whatsoever of him being a female. At the moment he sends his grandmother six or seven SMS texts every day saying, 'for the time being I have problems about being a trans'.

Another of Littman's findings drew attention to a pattern of transgender identification in friendship groups, revealing the possibility of social contagion. Clinicians and parents contributing to this chapter similarly report a localized nature of transgender activism, and its malleable potential can be seen in examples of the way children and young people discover their trans identity in schools, youth groups and friendship clusters.

Elsa's parents:

> At 17 years she came out to me as transgender. I was totally shocked as she never showed any signs of being a boy. Two of her childhood friends had come out as transgender and through immersing herself with happy trans stories they shared on YouTube and Reddit, she became a follower of the trans-cult.

Oscar's parents:

> Suddenly, at 15 Oscar said he was a girl. We were shocked since this was not something we had ever seen any signs of when he was growing up. We tried thinking what made him think this. For a few years he had been involved in the Student Council. When we found out more, we discovered this organization works to normalize the idea that people can change gender, sex and sexuality. The Student Council has embedded these ideas in his school. Oscar was searching for a way to belong and feel included and became active in internet transgender groups to do this.

Julia's parents:

> At her Eating Disorders Outpatients Clinic she met two transgender Female to Male individuals and became close friends with both. A few months later she started identifying herself with them. We found their messages supporting her and encouraging to 'keep going'. A couple of months later, she stated that she does not have any gender, and started asking for a short haircut. She even trimmed her eyelashes. I overheard her tell a transgender Male to Female classmate that she wanted to be a boy. That transgirl told her 'after getting hormone treatment you will feel so much better'.

Clinicians do not think the rapid rise in the number of children identifying as transgender in clusters is a product of new gender freedoms but the opposite:

> Transgender identification in groups shows freedoms shutting down because actually society has become more gendered in many ways. And linked with that, is the amount of normativity that we hear, incredibly regressive gender ideas which are used as the explanation about why a kid needs medical intervention, you know, 'why would a child need blockers?' 'Well they need blockers because they've always liked pink and that means he can't be a boy' or 'she wears jeans and always plays football so is really a boy'. But all of that, you can talk about that with colleagues who are gender critical allies but when it comes to clinical action those concerns are not accepted as something that could stand up as a reason not to proceed or act. (Anonymous gender clinician, 2018)

As Littman showed, parents know that a failure to explore ROGD leads to misdiagnosis of transgenderism and perilous neglect of the real issues underlying their child's gender distress. Clinicians from two separate world leading gender identity clinics confirm that an affirmative approach to transgender intervention means comorbidities are missed because a detrimental service distinction is structurally embedded into practice:

> ...the doctrine is to treat gender dysphoria and mental health problems as two independent issues. (Anonymous gender clinician, 2018)

> ...this means we are endorsing intervention on children's bodies without accounting for variability ... a tolerance to self-identification is extended which is actually repressive ... you don't just see one child and understand gender identity is not innate but once you've seen a hundred you've seen 'the reddit kid', you've seen 'the teenager with autism', 'the one who might be gay', you've seen 'the girl who was sexually abused and hates her body' or whose mother has been sexually abused and hates her body and

doesn't want the same for her child ... we know that by not examining what is behind the onset of dysphoria, and going straight for self-affirmation that the patient is transgender, we are subjugating children's clinical needs to an ideological position. (Anonymous gender clinician, 2018)

Like the reflective clinicians, what parents say about their children's self-identification as transgender consistently reveals it is a response to circumstances the child is finding difficult.

Anna's parents:

> The school had major problems with bullying in the girl's group and she was involved in this. She was so sensitive she would start crying in class. Her weight increased, she cut her hair and people said she looked like a boy. She became darker in mood. We begged gender clinicians to look at what Anna had been going through that might give insight about why she changed so rapidly and decided she was a boy. No-one had expected this; the girl, who loved dolls, dresses, glitters and ribbons, was no longer identified as 'she', but suddenly became a dark and gloomy 'he' binding her chest with duct tape. We were not heard. Anna did not get the help she needed to process possible trauma. Instead she was guided and supported into a new life as a boy, without any kind of reality orientation or care of her whole as a human being.

Anonymous clinician:

> Once the child announces that gender is the problem this masks all other issues. Referring clinicians assume that gender problems are too specialist for them to deal with and that the children they refer to the gender identity development service will then be receiving psychotherapeutic care. But that's not the case; once the child gets to the specialist service the clinical work is all about fast-tracking gender-affirmation and transition.

An extract from the account of Filip's parents indicates service level policy interest in fast tracking transition.

Filip's mother:

> He kept going to the transgender clinic because he wanted hormone pills prescribed. No initiative was taken to treat or refer him for ongoing depression. It is uncertain if his mental health was assessed and how it was taken into account in the investigation of his gender dysphoria.
>
> Within six months of receiving a diagnosis of gender dysphoria Filip has had an assessment of his larynx for eventual surgery and received an

appointment regarding genital surgery. He asked me to cancel these appointments saying he was not interested in any surgery. In the letter he received there was a document to be signed, to ensure that he had been fully informed and understood that the consequences of the surgery are irreversible. Filip told me he hadn't asked for any referrals and he thought they might be routine for people with a gender dysphoria diagnosis. It is of deep concern that Filip (and probably other young people around the world) are being sent from gender clinics to surgeons without any cast-iron certainty that they understand what is going on. He has repeatedly said he is not interested in having surgery.

In the meantime, Filip has no functioning everyday life. He has not informed any relatives except me, his mother, about his gender dysphoria. His clothing and appearance is not yet female. It isn't obvious that he would like to be female. He has not asked for us to use another name or pronoun when addressing him. Filip's father, whom I was divorced from 5 years ago, has very little knowledge about Filip's gender dysphoria and knows nothing about him taking hormones and all the other aspects of proposed intervention. All he knows is that Filip has thoughts about his gender identity.

Clinicians using a self-affirmation model of gender care position themselves as impervious to family insights. They are able to mobilize ideologically-based ideas upon which to base intervention, even to the extent of thereby renunciating other required medical care.

Julia's parents:

> We went to the Eating Disorder clinic because she was already diagnosed with anorexia at 12 years old, just after the menarche, when she started spending a lot of time on [the] internet and started restricting food intake. She had been hospitalized before because of vital parameters and now she was not stable in weight. But our daughter said to the doctor 'I am a healthy transgender, I don't need eating disorder treatment, refer me to [the] Gender Identity Center. I am a boy'. She was immediately referred to the children's gender identity clinic; the Eating Disorder clinic said they would not any longer treat her. I remember the psychologist told me that I have to accept she is a boy saying 'otherwise your child will relapse in anorexia or die'. We felt that the therapist was wrong.

Elsa's parents:

> She had just ended her first long-term relationship with a boy, with a hard break-up, ending in threats from the boyfriend that he would commit suicide if they did not get back together. We asked the clinician why she

was so disturbed ... they said 'because she is born in the wrong body of course'.

Tuomas's parents:

> Tuomas became psychotic and received sex-reassignment surgery despite our warnings that he was obviously heading into schizophrenia when he began identifying as transgender. We sent letters to clinicians and attempted to intervene via the health supervising authorities. No proper assessment of his socio-economic or psychiatric status was done.
>
> After the sex-reassignment surgery was done we pleaded with the hospital to give him proper treatment for his schizophrenia, this to no avail. He was treated solely for gender dysphoria.

An affirmation approach to transgender identification directly leads to deprivation of access to much-needed mental health services. Particular concern was identified by parents and clinicians that an affirmative model of transgender intervention does not offer an appropriate route of support for children with pre-existing autistic spectrum diagnosis, again endorsing a finding reported by Littman.

Oscar's parents:

> Growing up, he had a hard time understanding social codes which made him feel he wasn't like everyone else. When he was about 10 or 11 years old, he received a diagnosis of ADHD with some autistic traits. At that time, he also had thoughts about his body and his appearance, he was dissatisfied with his body. As is typical for Oscar with his Asperger's traits, now he has decided he is a girl, no one can change his mind. He will not listen to any other points of view. He has started to take SSRI medication but is not happier or released by his gender transition. I don't think changing gender is solving any autistic spectrum problems for him.

Elsa's parents:

> She had social and communication difficulties in school. She was sensitive to sounds and her emotions have always been strong and rapidly shifting. She has little sense of time and finds it hard to organize things and to concentrate on things that do not interest her. When she is interested in something she gives it her all, and during that time, she is convinced that this is what she wants to do for the rest of her life. The interests are intense, but the focus of interest differs.

Filip's mother:

> I am concerned that research is needed to properly understand ROGD because taking a blanket approach to affirmation of transgender identification has meant that Filip's autism has been completely overlooked to produce a clinical picture that sanctions sex-reassignment without any attention to his autism. For Filip, and other young people with similar social problems, autism or autistic traits, transgender identification seems to be an expression of other underlying problems.

Clinicians agree:

> For those with autism and gender dysphoria intervention all becomes organised around the gender ... everything is storied as being around the gender. (Anonymous gender clinician, 2018)

The effect of the affirmative approach to transgender identification is to restrict the allocation of understanding and access to a range of support services:

> Cases that are worrying involve children whose family history is not explored, histories that include abuse, autism, clearly homophobic families, examples of survivor guilt where a child in the family had died and it seemed as if the cross-sex identification might be a response to the loss of the sibling. We see young girls internalising misogyny but attempts to hold a space open for reflection and exploration is seen as elitist and patriarchal. (Anonymous gender clinician, 2018)

All of these issues require a space to be held open for thinking through the multiple and complex personal, social and ethical dimensions of gender reassignment. Yet clinicians say they are compromised in their ability to help children process the reality of transitioning. An issue that particularly illuminates the importance of discussion for avoidance of great harm is that of future fertility:

> I feel there's something really dishonest about the effort going in to getting children to preserve their fertility. What are we setting them up for? It's not enough to talk about fertility preservation. We are playing down the reality way too much; for example, we can't not think properly about the process of getting a child to go and masturbate in an IVF clinic. Should we gloss over questions about whether children with massively traumatic histories will really be able to adopt? We aren't talking enough the reality of any blocker or hormone treatment massively reducing the chances of them being able to preserve sperm or eggs. It would be a more honest conversation to say 'you are almost certainly sacrificing having children'

but the demands being placed on us are for not thinking. (Anonymous gender clinician, 2018)

To sum up, the affirmation approach to transgender identification precludes the exploration of the origins or significance of gender uncertainty. Prevailing models of gender intervention view a gender-distressed child or young person within narrow, ideologically-driven limits and deny them access to any other paradigm of support they may need.

Contesting the exclusion of parent perspectives on transgender identification

Again, matching evidence reported by Littman, accounts from families in Scandinavia reveal that preoccupation with self-affirmation in models of transgender care automatically positions parents as subject to criticism, poorly informed, bigoted and not operating in their child's best interest if they ask for more profound attempts to make sense of their child's transgender identification. Self-affirmation as the driving principle on which intervention is based completely takes away the power of parental perspectives, as some gender clinicians realise: *'I know there are parents and families desperate for a different model'*. Parents elucidated on this.

Anna's parents:

> We thought we would be in the safe hands of specialist gender clinicians who could explore the past, uncover trauma, and help 13-year-old Anna to process her confusion and distress. We were confident they would not automatically comply with a child's demands for treatment with testosterone, for beard growth, a deeper voice and breast removal without any exploration of what had happened to suddenly change her. But we were wrong.

Parents who are concerned about medical transition or who question the affirmation approach are positioned as 'unsupportive parents':

> We were told we as parents needed help, not Anna. We were assigned our own family counsellor who guided us to come to terms with Anna's sex-change. Anna was assigned her own psychologist. According to the psychologist, what Anna has said is confidential. The outcome created the gender clinic was that the distance between us and Anna was assured: the service had created a situation where we as parents were treated as adversaries to our own child. This insistence on making Anna into Jaakko could not be right—or was it? The Gender Clinic insisted it was.

I felt that the world shivered.

Just as transgender-identifying children are encouraged online to view their parents with suspicion as 'transphobic bigots' if they question self-affirmation, they can be encouraged to alienate themselves from family members by medical practitioners too.

Tuomas's parents:

> Encouraged by the gender doctors, he decided to cut off contact with us.

Parents question the legitimacy of their exclusion.

Filip's parents:

> Close relatives were never involved in the investigation even though this is highly recommended for every other situation involving distressed young people; gender dysphoria should not be managed differently, ignoring family insights makes for suboptimal care.

They lucidly explain the reasons why they seek to avoid pathologization and hold back lifelong decisions being made about intervention without evidence and without their input.

Anna's parents:

> We just want Anna to wait and see how she feels about her sex and gender as she grows up. At 13 she does not have the prerequisite experience to understanding the consequences of choices being made now. We think it is terrifying that the specialist clinicians assume the authority of confirming reassignment of sex and gender for children at such a young age without involving parents and with no acknowledgement of how much the internet affects the current cohort of children identifying as transgender in unprecedented numbers. Parents like us are marked out as conservative, evil, prejudiced and accused of destroying our children. We are not those things. We are an ordinary average family. We support our child and have a good relationship with her.

Not being involved in discussions of ROGD leaves parents and children and young people vulnerable.

Elsa's mother:

> I started to read up on the science around the subject and was horrified by the lack of evidence for the affirmative approach treatment these young people (even children) are given but the clinic won't answer my questions. My daughter has now been taking testosterone for approximately 8 months and has a double mastectomy booked. She lives with her homosexual boyfriend. My gut feeling is that a sex-change is not the solution to her unhappiness.
>
> I'm depleted of energy and hope for Elsa's safe future. We only have sporadic contact with her and to have a relationship at all we cannot bring up anything around the trans issue.
>
> And throughout this entire episode, society tells me that my strong feeling of protecting my daughter is wrong. And clinicians won't look at research evidence.
>
> I have told Elsa that I will fight the so-called specialist gender clinic until the day I die, because they are experimenting on troubled young people like her.

Anna's parents:

> We now have no doubt gender change is not right for our Anna. We had a healthy and normal girl, with normal development, who suddenly changed behavior after bullying in school. Then isolated with a smartphone she came and told us she's a boy. Should we follow the advice of gender clinicians and not attempt to process anything of what has happened? Should we only comply with her new name, new pronoun, and get testosterone into a healthy body? At the age of 13? In the middle of puberty? At an age where everything is already in chaos before we contemplate adding in a gender shift with no evidence of healthy consequences? For us as parents this intervention strategy is unconscionable.
>
> We are shocked that children's services threaten to lock Anna in to a decision to change sex at the age of thirteen when she has so much developing yet to do. What happens if she changes sex and gender and then changes her mind? What kind of provision is on offer then? After years of convincing yourself and insisting to others that you are the opposite sex, how can a young person stand up and say 'I was wrong. I surely am the gender I was born as'. How easy is it for a young person to contradict the service providers who have colluded in their sex change? What scars will this leave on the young person's body and mind? How then can they live on?
>
> Anna clings to the idea that she is born in the wrong body. She needs time and maturity to understand that she has been manipulated by the trans-community. Clinicians tell us she was 'born in the wrong body'. This is clearly an invention of their own.

Tuomas's parents:

> We had tried to stop the progress of the reassignment surgery, we tried and tried before the surgery to get the doctors to instead give him psychiatric care. After the surgery his mental health deteriorated further and he became chronically psychotic. He claimed to be an Indian Goddess. He was 28 when the sex-reassignment surgery began and lived until 32.
> After repeated deeply psychotic episodes involving repeated police interventions and forced care a few months following the first sex-reassignment surgery (which was to be followed by additional prescribed surgeries) Tuomas committed suicide in July 2017. Our interpretation is that a few weeks after he was stabilized following sex-reassignment surgery he realized the full consequences of what he had done—what had been done to him—and decided to end his life.

The need parents have to be involved in the care of their children is sharpened by the knowledge they have accrued about clinicians operating without an evidenced-based model and resisting evidence.

Anna's parents:

> We contacted the psychologist and sent research papers about alternative diagnoses we found including ROGD, but we were simply told that it was important that we supported Jaakko in his process. We were left with a huge sense of powerlessness.

Elsa's parents:

> We took papers reporting scientific data that gender confused adolescents are not better after their sex change to meetings with Elsa's counsellor. The counsellor did not accept this evidence and just told as 'truth' that 'almost nobody regrets transitioning' and that gender dysphoric children kill themselves. Moreover, she said my feeling of distance towards my child if I was not able to call her by masculine pronouns was 'wrong'.
> I asked what they were treating my child for to which the clinician replied—'Ok if you want it that way, we are treating her for testosterone deficiency, since she says she is a man'. Moreover, she told me that the way they use cross-sex hormones off label is 'the same for every drug in the world'. Having worked with drug safety assessment for many years I knew this claim to be false.
> At every meeting the clinicians brought up the high rate of suicide among gender dysphoric young people which I interpreted as a threat to make us quietly comply with transgender ideology.
> Several remarkable claims from the so-called specialists have been expressed at these meetings when I have argued that these young girls are a

new and unique patient group that they know nothing about, and that their treatment therefore must be considered experimental.

They defend their insistence on self-affirmation and denial of ROGD by saying:
- This is new ground being broken.
- The question is not about the risk, but rather a great suffering that needs treatment.
- No information on history from the family is needed for the persons over 18 (even though we told them that we know Elsa lied on the self-declaration form).
- Those with problems to begin with will still have those problems after transitioning 'That's life!' they said with not a trace of awareness of implied ineffectual 'treatment'

And they don't give any written information to their patients.

The position and claims of gender practitioners are clinically unintelligible. Below is the evidence that self-affirmation of transgender identification is unreliable.

Julia's parents:

February 2018
Recently social services decided to take her from us, placing her with [a] foster family for 3 weeks, to investigate our family. While placed with the foster family, she relapsed into eating disorders. The social services compelled us to use male pronouns and name, buy her boys briefs, and let her have a mohawk-style haircut. It took several months of 2-3 hour weekly visits from social workers to realize that we do care about our child very much and get her back. Now she is 15 years old and at a stage where she is painfully aware that she can never become a real boy despite names or hormones or operations. She is convinced she is a boy and that she is a prisoner in a wrong body. She has been transgender now for 2 years and 3 months, uses male name and pronouns. She is using a breast binder, stuffing briefs with rolled socks to imitate a presence of penis. She is in contact with transactivists, she twice ran away twice with transactivist youths from the support group. Unfortunately, after social transition, she does not feel better. She has a lot of anxiety, mood swings, irritability, she is very tired. She is difficult to reach, and she is completely different from how we knew her from the first 12 years of her life. The GID clinic are going to finalize their observations, recommending treatment for her. I don't know if I will ever have my child back.

February 2019

She has broken away from the transactivist youth group and doesn't want to go back to the GID clinic. She said to me 'Mummy, they are crazy. I can't turn into a boy. I want to get my healthy body back'.

Co-production of knowledge about transgender children and young people

Critics of ROGD will disavow the validity of stories from a small number of parents, but such criticism only betrays confusion about the importance of internal versus external validity. All insider perspectives have their own internal validity and clinicians know these are vital for strengthening intervention guidelines and practice. The diminishing of parent perspectives by those intent on muzzling experiential knowledge and evidence which does not fit with their own predispositions cannot go unchallenged. Activists cannot be permitted to dispense ideology to mobilize clinical interventions for children and young people who experience gender dysphoria. Parents have important knowledge too which must inform change. Littman's data also support the inclusion of parent perspectives in the diagnosis of their child to balance reliance on self-identification. This is not to say parents should be exclusively responsible for confirming or refuting their child's self-identification as transgender, but that the current practice of excluding parents from the clinical picture collaborates in the invention of the transgender child and in actively promoting and reinforcing transgenderism. The tension between sets of commentators is acute, in that ideologies fuelling the provision of medical care to children and young people are not evidence-based and seek to block the emergence and dissemination of information parents hold. Knowledge and expertise that comes from the experience of parents must be properly valued by the research and clinical community, not suppressed. Of course, no claims are made about the 'representativeness' of these stories. Their authenticity, however, speaks for itself. Those who sent in their stories are all concerned about gender transition, uneasy about the unquestioning affirmation approach, or simply do not agree with the unevidenced hypothesis of innate gender identity, but they are not, as can plainly be seen, 'unsupportive parents', as their opponents make them out to be. Their stories demonstrate the power of personal narratives for dismantling the invention of the 'transgender child'—which is undoubtedly why lobbyists wish to silence or devalue them.

Conclusions

The stories are examples of how free-flowing, unregulated claims about transgenderism have interrupted specialist services and effectively taken away any capacity, authority or experience clinicians can bring to bear on understanding individual children and young people's gender dysphoria. Clinicians are operating in a context which emerges from constraints upon engagement with research and literal attacks on attempts to build research-led, evidence-based practice that calls for an understanding of origins of gender dysphoria or for the involvement of parents who might want to ask questions. Demands of transgender activists are so entrenched in the culture and mind-set of specialist gender identity development services that the reality of ROGD, and its exposure to the abnegation of care that is embedded in acceptance of gender identity, are being ignored. It is as if there is a collective unconsciousness amongst clinicians treating children and young people who think they are 'transgender'; a collective disassociation led by professionals, from the fact that a person cannot be born 'in the wrong body'. The usual boundaries of evidence-based medical practice and the knowledge and protection of parents have been turned into no-go zones, leaving the different unregulated influences of ideology and activism to take over without constraint. We are giving rise to a global network of GIDS which insist on remaking bodies with rubber penises, wounded breasts and unsafe hormone treatments, allegedly to eradicate gender confusion rather than explore and deal with gender-based harm.

This chapter can only provide a brief review of some of the dangers of suppressing ROGD research. However, the parent and clinician perspectives brought to shed light on the phenomenon in this chapter show it will be important not to neglect research in this area or to reject emergent findings. Disruptions to research which necessitate responding to journalists and spending hours, days and months refuting baseless accusations are stressful, but they highlight the imperative for a full review of patient experience, including the perspectives of parents whose input can inform treatment recommendations.

Clearly, when entering the world of transgender identification, there is a danger of disengaging from the real causes of gender distress. The data in this chapter—as across the entire book—guard against transgender identity being seized upon to neutralize all other concerns about a gender dysphoric child or young person. A matter of utmost importance is the complexity of transgender identification which has been made visible through the parents' accounts, inescapably showing that children and young people rarely express the totally certain self-diagnoses upon which affirmation-led intervention is based.

This brings us to focus, towards the close of this book, on some of the ways in which transgender intervention can be improved. Parents provide clarity about the voices which most need to be heard to improve transgender policy and practice.

Anna's mother:

> I am concerned that an affirmative-model of intervention means ongoing mental illness is neither treated nor is considered before giving treatment for gender dysphoria. I wish that researchers and the authorities would listen to all the young people who decide they are transgender out of the blue and to all those who now say they regret their transition.

Clinicians are increasingly telling us they are meeting young adults who have transitioned, massively regret it and then detransitioned, and detransitioners are speaking out with growing confidence, again as we have seen in this book. Clinicians are, however, concerned about long-term follow-up of the current exponentially growing cohort of children and young people:

> Those transitioning doesn't work out for will fall off the radar. They won't want to come back and speak to us because it is all too painful and embarrassing. (Anonymous clinician)

The outcomes are not predicted to be good:

> My prediction at this point is there'll be three groups. There'll be a group who are OK because humans are incredibly adaptive and they will survive and they will make the best of it. I think there'll be a group for who it wasn't OK but they cannot face the fact that it wasn't OK and they will be lobbyists of the future, unhappy and miserable but not able to locate their unhappiness in their detransition because that would be too much of a cost. And I think the third lot are the detransitioners. The detransitioners are the healthier version of the future lobbyists. And there's a fourth category who are just long-term struggling people who are emotionally unstable.
>
> I mean genuinely, when you think of children who transitioned at 5 or even 15 where do you start unpicking that? Where do you start? (Anonymous clinician)

Anna's parents:

> We will always be by our daughter's side supporting her. Perhaps our story can help inform treatments for children and young ones who are confused about gender. Parents are powerful and finding each other—we are no longer alone.
> My prayer goes to gender practitioners; it is imperative to make sure that children and young people get help to process gender confusion BEFORE they are prescribed untested medications and impelled to flee from their own body.

You have read this book. Now, where do you start?

References

Brunskell-Evans,H. (2018, 11 December). Rosa hjärtan i blå kroppar: Barnen på operationsbordet. *Kvartal*. Retrieved from https://kvartal.se/artiklar/rosa-hjartan-bla-kroppar/

Littman, L. (2018). Rapid-onset gender dysphoria in adolescents and young adults: A study of parental reports. *PlosOne, 13*(8), e0202330. doi:10.1371/journal.pone.0202330

—. (2019). Correction: Parent reports of adolescents and young adults perceived to show signs of a rapid onset of gender dysphoria. *PlosOne, 14*(3), e0214157.
https://doi.org/10.1371/journal.pone.0214157